THE
DIGITAL
HEALTH
REVOLUTION

THE

DIGITAL HEALTH REVOLUTION

KEVIN PEREAU

TranscendIT Health

Walnut Creek, CA

Paperback ISBN: 978-0-578-40972-6

eBook ISBN: 978-0-578-43145-1

Book design by *the*BookDesigners

To Beth

ACKNOWLEDGMENTS

First and foremost, I want to express my love and gratitude to my wife Beth. I often tell people that she knows more than anyone does about health care, because that is absolutely and completely true. She came to it because she is someone who cares deeply about people and miraculously, she cares about me too. I also want to express my love and thanks to our wonderful daughter Megan, who also caught the health care "bug" and who now works as a consultant in the field. Without the support of Beth and Megan, this book would never have happened.

I express profound gratitude to all the experts who contributed their views to this book, enriching it miles beyond what it would have been without them. They took what was an interesting book and made it exceptional.

I would also like to thank just about everyone I have ever worked with over the course of my career. In their own ways, each of them shaped and changed me as a health care thinker and professional. And let me give special thanks to Rick Lee, my collaborator and co-founder of TranscendIT Health.

I thank Barry Lenson, the writer and editor who helped me turn a lot of ideas and plans into this book. We had fun and became good friends in the process, and I thank him for that.

Last, I would like to thank you, my readers. You and I will enter an amazing new world of connected care in the years ahead, an amazing and healthy journey we will take together. Thank you for joining me, and welcome aboard!

CONTENTS

FOREWORD

By Penny Moore

Partner, Commonwealth Health Advisors

Digital Health Enthusiast

Stop for a moment and think about the mobile apps you access on your smartphone more than five times a day. Text messaging. Email. LinkedIn. Facebook. CNN Newsfeed. Would you say these mobile apps are embedded into your daily routine? Would you say they have impacted your approach to life? No doubt, they have.

Now imagine a program designed to help individuals with chronic conditions change their lifestyle habits and get healthier by sustaining an average of 5.2 daily interactions with its participants over 16 weeks. Impossible, you say?

It is happening! I have seen it.

The *Digital Health Revolution* explores how progressive technology advancements are democratizing health care for the consumer. New digital technologies, big data analytics, and sophisticated AI are now innovating health care delivery with powerful digital apps. It is the contemporary creative digital design that draws an individual in to easily consume clinical evidence-based medicine in actionable methods that naturally become part of daily living.

I am a seasoned health behavioral change skeptic! The hardest thing to do is to get people with chronic conditions from lifestyle etiologies to change their behavior. Even people with the best of intentions fail at sustaining healthy behaviors.

For more than twenty years, I have dedicated my professional life to the evolution of population health and disease management. My passion has been driven by the sobering realization that during the last century we were dying from diseases and viruses we caught when we left our homes. Now, it is from things we do to ourselves. Our daily lifestyle choices matter, and they all add up over time.

Since 2011, I have immersed my focus on the advancement of behavioral change programs by the digital revolution. I am awestruck by the exponential improvement that consumer-centric digital design, machine learning, and conversational bots can have on achieving what clinicians, behavioral scientists, and others have only imagined for over twenty years.

For over three decades, physicians have been taught that the first line of therapy for type 2 diabetes and early stage cardiovascular disease is diet and exercise. It seems physicians have given up on people changing their lifestyle contributors to disease. Many societal and health care trends have played into that. It is much easier for physician and patient to utilize drugs to manage these chronic conditions and slow their progression.

Current digital and technological advances are making it possible for lifestyle behavioral change to not be dependent solely on individual *willpower*. Today's digital tools, leveraging modern data application, are empowering consumers to put *skillpower* to work changing their lives. To sustain behavioral change, people need consistent input. They need the right information, timed to a meaningful, teachable moment, suggesting actions based on their own data to build skills and routines that shift their daily approach to living. Consumers are quickly catching on that these digital tools are creating sustained changes in their lives that are relieving them from the burden of a life sentence of taking drugs to stay healthy.

So, plug in - get connected to *The Digital Health Revolution* and benefit from the insights of how health care is now transforming from physician-driven care to empowered patient-physician partnerships.

ABOUT PENNY MOORE

Penny Moore is a partner with Commonwealth Health Advisors. She has devoted her career to providing leadership to growth-oriented companies that are transforming population health through innovation with digital technology. Most recently, Penny has led the commercialization for early stage digital health start-ups Kurbo Health and Better Therapeutics. As the Chief Revenue Officer for ShapeUp, a venture-backed company that pioneered social wellness, Penny's leadership created a high-quality book of global business opportunities that paved the path to a successful exit. She has held executive leadership roles with large health plans such as Kaiser Permanente, UnitedHealthcare and Aetna.

INTRODUCTION

When I was a kid growing up in a rural New England farming community, my mom always had her hands full. My two sisters Patricia and Karen had already pretty much grown up and moved out, but three boys remained, constantly challenging one another and pushing against every boundary we encountered. My older brother Bob was the rebel. He was the reason rules existed. He pounded on us mercilessly to the point where my younger brother and I had no choice but to band together in mutual protection. We learned early that to be fast meant not getting pounded. My mom's time-honored method for refereeing the indoor skirmishes was constantly telling us to go outside and play. That's exactly what we did.

If you have never spent much time in the country, you probably have little appreciation for how you keep yourself entertained there. We had no cinemas or shopping malls. We rode our bikes around town looking for enough kids to play baseball, basketball or football, depending on the season. Not all of us were into sports and when you grow up in a town of about 800 residents, you learn to diversify your activities. Fishing was probably the most popular. We would pedal our bikes down to Norton's Hardware in downtown Fair Haven, Vermont, purchase just enough bait for the afternoon and let old Ralph Norton check our poles to make sure everything was in working order. Then off we would ride.

Mendon Pond was about five miles outside of town and to get there we had to either take a busy two-lane highway called 22a, which connected the thriving metropolitan area of Fair Haven to a point just south of Burlington. My dad used to call 22a the "Ass" highway because when you looked at 22a in the mirror, that's what it spelled. I still think that's funny. What made it seem like a busy road even by Vermont standards was that it was winding and narrow. It always had big freight trucks on it which sometimes honked or swerved at the kids on bikes to scare them. Sick and twisted, I know, but hey, welcome to Vermont. We typically opted for the dirt roads that jutted up one side of Rattlesnake Ridge, cutting across the orchards and slate quarries that brought early settlers this far up north back in the day. Of course, going the back way made it a longer haul and the dirt roads meant constantly dodging potholes and rocks big enough to send you flying over the handlebars. On good days, we would bring home the booty. Nobody needed to tell us how to catch and clean fish. It was hardwired into the DNA of every kid, big or small, growing up in Fair Haven. On bad days, we would raid Sheldon's Orchard and bring home just enough Macintosh apples for our moms to bake cakes and turnovers.

With every passing season came a new activity. New England isn't for the faint of heart. Those seeing the fall colors and thinking, "What a lovely place," have never shoveled snow. Sure, they see it coming down and think it's beautiful...the first couple times. Those of us who grew up there knew better. In the winter, the boys' job was to shovel out the driveway. I always got out there first and took the part close to the garage. It was where the basketball hoops were located, and my incentive was being able to shoot hoops all year long. I was meticulous about how clear I kept it. I mean, not a bit of snow was ever left on this part of the driveway. My dad was always proud. I took the largest section and cared for it like a drill sergeant on new recruits.

My dad happily bought me about three or four rubber basketballs as a reward. He knew how to keep me motivated. In the winter when the temperatures dropped, I would store them near the radiators. If you've ever visited a home built in the 1800's, you've seen these big monstrous-looking metal things in every room. They dominated the décor. You don't have to be a Tom Brady fan to know what happens to the air pressure in anything inflated when the temperatures drop. I shot hoops while my brothers did the rest of the driveway. When the ball lost its bounce, I would simply go inside and get another one and put my Brady ball back near the radiator. I would repeat this while watching my brothers finish the driveway.

I always took the largest section because it was the easiest to do. My older brother Bob took the middle. It was the smallest section, and he thought he was getting a good deal. Heck, he would have made our lives miserable if he thought otherwise. Younger brother Dave always got the shortest part of the driveway, and the shortest end of the stick. He got the part closest to the road. It was all good until the snowplows came by to clear the street. To this day, I still hear stories from my neighbors who could hear from three blocks away my dad bellowing, "David Michael Pereau—get out here right now!" I don't think my younger brother has forgotten or forgiven me to this day ☺.

In the springtime, it was planting season. Since I never took to hunting in the fall, this was one special activity that I did with my father. I was a gym rat growing up. There wasn't a sport I didn't play, and my dad was always there for me. I was playing catch, shooting baskets or throwing the football. I always wore the old man out, and helping him plant and weed the garden was the least I could do. It was our quality time. He got plenty of that with Bob and Dave during hunting season. He and I spent many hours out in our back yard and he let me plant anything I wanted. From strawberries to rhubarb to pumpkins, I had my own section, but I had to care for it all season long. Our neighbor had the biggest patch of raspberries, blackberries and blueberries I had ever seen. We were never hurting for munchies. We just went out back or next door and ate until we were full. Food was fuel to a growing boy, and we were constantly on the move.

Come summer time, my brothers and I would scrape our nickels and dimes together until we had just enough money to rent a boat on Lake Bomoseen. We were too young to rent anything with a motor, so we would row out to the island. It was famous for decades as a retreat for Hollywood stars like the Marx Brothers, Marylyn Monroe and others. Believe me, there was quite the ambiance when eating out in beautiful downtown Fair Haven. Truthfully, we didn't care about any of that. Years earlier, the older kids had tied a rope to a tree overhanging the cliffs. We rowed our boat out to jump off those cliffs. We were too scared to dive, so we all just jumped in feet first. Anyone boasting about doing any different was lying. Today, there is an eagle's nest on the rocks we used to jump off. The tire swing is long gone, as are the Hollywood stars who used to frequent the Algonquin Club located on the island.

By autumn, we were in full leaf-raking mode. There were no blowers back then, and no sacks for stuffing the leaves for a later pick up. This was pre-environmental anything, so we burned the leaves on the side of the road in the space between the blacktop and the hunks of slab used as sidewalks. They were too imperfect for anything else, and I still remember the horses and trailers hauling them up from the slate-cutting shops and laying them down along our streets. Nothing says New England like shuffling through the leaves on a slate sidewalk. When my mom was overcome with debilitating dementia and Alzheimer's disease, I would walk her to the corner and back. She felt like a kid again.

Looking back, I wouldn't change a thing about how I grew up. We were never bored. We had no smart phones, no electronic gadgets of any kind, and we only played indoor games when it rained. We were active, we ate healthy foods and we participated in a variety of activities that kept us moving and kept life interesting.

We resided at the original corner of happy and healthy. Heck, we probably inspired that as a Walgreen tagline.

Sitting in a health care conference years later listening to the speakers drone on about chronic conditions and how inefficiencies in our health care delivery system were contributing to chronic conditions that were overwhelming our health care capabilities, my mind wandered back to my childhood. I had the mother of all epiphanies:

We can absolutely fix our broken health care system . . .
and I knew exactly how to do it.

CHAPTER ONE

OUR SYSTEM IS BROKEN, BUT WHO CARES?

Have you ever had a friend who constantly asked why things are done the way they are done? I was definitely that kid. When I first shifted my professional focus from technology and management consulting to health care, I was a nonstop "But Why?" machine.

I must have driven my poor wife crazy. She has logged more than 30 years in the health care industry, where she came up through the ranks. There isn't an aspect of how insurers intersect with our lives that she can't go into the weeds to discuss. She started in underwriting, later moved into account management, then did sales and before long, moved into management and was running regions. Okay, I glossed over quite a bit there because she is a woman and it took her twice as long and she had to work twice as hard as her male counterparts to finally become president of a multibillion-dollar region, and then CEO. If her name were Bob, we would probably be living on an island already. But you get the picture. She knows the health care industry backwards and forwards, inside and out. Forget the fact that she is only one of a handful of women in leadership roles. That is a story for another day. She has given me an insider's perspective into all things health care and C-level connections any digital health startup would covet. From health care consumption to servicing, I have seen it all—and it isn't always pretty.

AN EARLY VENTURE I MADE IN HEALTH CARE

When some former tech colleagues approached me about an idea they had about starting a health scoring company, I was intrigued. I had long asserted to my wife that one of the reasons we have such runaway costs in health care was that nobody measures anything.

How can you manage something you don't measure? Seems like a logical question.

In coming up with a way to measure an individual's health, we borrowed a concept from the financial industry and applied it to health care. We started with a simple score, something that anyone could track and understand. It was amazing how closely it tracked like a credit score. We used a sliding 0-1000 scale. At zero, well, you were probably dead. At 1000, you were taking the leading role in Superman movies. In between, anything above 600 was good; about 700, excellent; and above 800, exemplary. We put an arrow by your score to show whether you were trending up or down.

Yet there are some significant differences between your health and your credit score. For starters, your credit score only tells half the story. It shows how much you owe and how well you service your debt. It doesn't list wholly owned assets or how much money you make. How do you account for similar factors when expressing the status of somebody's health? We knew that would be challenging, and realized we had to get it right. After all, you aren't automatically healthy the day after you quit smoking, just as you aren't at risk for a heart attack or diabetes because you skipped a day of working out.

Starting in 2011, a cornucopia of data-collecting devices, ranging from the Fitbit to the Nike Fuelband, were already exploding onto the scene. Withings had introduced scales that could capture and store your weight, body fat and BMI for later analysis. Welcome to the Quantified Self movement. We decided to be agnostic about what devices we would use to fuel the data that would drive your score. Garmin, TomTom, even your phone, were just fine. Your score would reflect who you were, what you did, and how you felt. We measured incrementally how daily lifestyle choices affected overall health.

We were going to change health care. Measure it, manage it — thrive. Boy, did we have a lot to learn.

While we got much of it right, we didn't see the incredible headwinds coming, coming from the entrenched incumbents who profited from fixing folk when they became broken. We quickly realized that not many in the industry were financially incented to keep people healthy. Sure, they talked a great game, but at the end of the day if we didn't get sick, nobody made any money. Not the doctor anyways. How could we get their oars rowing in the same direction if it meant cannibalizing their core profit centers?

The first pushback from early innovators in digital health came from the insurers. When presented with a value proposition that engaging the consumer would alleviate strain on the entire system, they responded with a consensus . . .

"Good luck, we have tried. Consumers are the problem. They won't engage, they won't share, and they won't take responsibility for their own health."

That, I thought, was a total sack of hooey. With my typical energy and enthusiasm, I dove headfirst into fixing health care. We would show the world how to engage consumers, capture and analyze data to better understand what was really causing our health care problems. We were pioneers in the digital health movement.

We thought it all started by getting consumers to know their health score. If we could engage people and get them to sustain their interest in taking better care of themselves, we were on the right path.

We learned incredibly insight-filled lessons along the way. While it's a fair call to say that all people are different, it is equally fair to say that there are many things we share. For sure, there will be a stubborn demographic that simply will not engage. No amount of coaxing seems to work—or does it?

One simple truth? We are all social, and we all seem to respond when gaming principles are properly applied.

ADDRESSING THE BIGGER ISSUES

Let's focus on how to best leverage social principles first. When we first came to market, we were looking for any path to the consumer we could find. We thought selling through partnerships with insurance companies would accelerate our market adoption. Insurance companies would sell health scoring through their broker channel to employers who would provide it to their employees as a benefit.

A whole new lexicon developed around this very convoluted and, as it turned out, challenging route to market. The business-to-business-to-consumers model (B2B2C) meant our business would sell to your business, who in turn would provide it to the consumer. Per Employee Per Month pricing became the norm. What PEPM are you getting? Questions like that dominated the conversation at health care tradeshows.

A health score alone was never going to be compelling enough by itself to get anyone to bite, so we reinvented our messaging. Since we allowed our users to track more than 100 activities, we developed ways to create challenges around anything that can be tracked with a device.

We then fit into an emerging category called the "Wellness Market." Our elevator pitch was now about a health score that utilized a B2B2C business model and heavily leveraged social and gaming principles to influence people toward healthier lifestyle choices—whew, what a mouthful! To our delight, it proved right on the money. Companies were biting. They were implementing our health score and developing corporate challenges for walking, running, cycling and burning the most Metabolic Equivalent of Task, or, METs, in a month. If you could track it, you could harness a rewards program to it.

While pretty exciting, I thought we could do better. Early push back from employees was, "Oh great, big brother is watching." I realized that our co-workers might not be the best source of social agitation for keeping our interest and sustaining engagement. Maybe my co-worker doesn't like me? Maybe she wants my job? Maybe he resents our VP of HR being able to look over his shoulder at his non-work activities as much as I do?

Maybe those employees weren't accepting of our company's intent to provide a tool that their employer could use to watch them at all times. Hey, argue all you want that it's not fair, or that any company with enough forward-looking vision to be investing in you like this should be applauded. But I promise you that people are skeptical, and you can't change human nature.

I thought back to some early lessons I had learned in the tech sector. At Cambridge Technology Partners, we did something innovative and daring. We created the Cambridge Information Network (CIN). It was a website created specifically and solely for CEOs and CIOs. We created something we didn't expect—we created a sense of community. With that came incredible insight into how the executives we were selling management and technology consulting to thought about their problems. We never let our sales teams anywhere near the site, and we were always careful about applying lessons learned from what was shared on it. In other words, it wasn't a lead-generation tactic.

We developed what we called a Lyceum Program around CIN, run by one of the tech sector's consummate thought leaders, Thornton May. For one glorious summer, I got to work with him hosting lyceums all over the USA. For those of you who don't speak Greek, lyceum is a place of learning. It was eye-popping. We never once presented those attending with corporate overviews or followed up about specific problems they had shared with us. Instead, we would pack the room with CEOs and CIOs who would bare their souls to one another about what was and wasn't working in aligning technology with their corporate missions. I was flabbergasted. Here was the CIO of Praxair sharing with other executives the things I had flown about 1,000 miles to specifically discuss with him. In a safe environment, he was telling others about how he had failed, and was asking them for ideas on how to right his wrongs. His peers were doing the same. Collectively, they were banding together to solve the most difficult of their problems.

Poor Thornton probably put on about 40 pounds that summer zig-zagging across the USA, eating very bad chicken dinners while form-fitting his bottom into seat 4E on Delta. Why am I telling you about this? Because as you apply these lessons learned to digital health, you realize a couple of things.

One is that in a safe setting, people share. They open up. Maybe interacting with co-workers isn't the best venue for helping people feel safe enough to share? As if my co-worker could secretly give a bloody rip about my blood pressure and body weight - who does? I'll tell you who does care: your spouse cares, your children care, your family and friends care. If you only remember one thing, please let it be this. It is worth the price of admission. Whether you are managing a chronic condition or you are active and healthy, you can derive great benefit from tools that allow you to build your own social network.

I mentioned there were a couple of lessons learned. Here is the other one. People behave differently when they think someone is watching, and that is especially true if that someone is somebody they care about.

We developed a tagline, "It's Better Together," and used it in all our product messaging. Incredibly, the early blowback we got from employers was amazing. "We don't want spouses involved in our employees' health challenges," they told us. "These are private challenges and outsiders aren't eligible to participate."

Really? We just showed you how to influence your employees to be more active, and now you don't want to do it? Welcome to the early innovators' dilemma. We had to do controlled pilots to demonstrate that employees would be more responsive to health care challenges when they were working with their spouses than they would be if they were working with their co-workers.

Spouses or co-workers—guess which group responded better? This is what I like to call the Cheers Bar principle in action. Blasphemous, I know, but you are more likely to show up and use a tool in a setting where everybody knows your name.

Another lesson learned had to do with gaming. We originally thought that to leverage gaming principles, we would constantly have to do challenges that offered rewards. There is nothing wrong with that. In fact, it helps keep things fresh.

My most memorable challenge was a walking challenge I entered. There were participants from Europe, Asia, South America and the United States. I targeted a goal of 1,200 miles walked for the year. That averaged out to about 100 miles a month. It was a bit aggressive, but I thought doable. I won the challenge but fell a little short of my goal. I think my total for the year was 1,143 miles walked on four continents. My nearest challenger was a 73-year-old professor from Switzerland who doesn't like to finish second in anything. I realized when tabulating the results at year-end that a 60 year old and a 73 year old man finished first and second in the challenge. So much for our demographic not representing!

A couple of surprising things about being in this challenge remain with me. One was how supportive we old fellows were of one another. Judging from the smack talking from some of the women in the challenge and the younger fellows as well as the regional participants, this wasn't uncommon. Support groups formed along natural lines and provided motivation to help us keep one another going. My nearest competitor would sometimes ping me when I was inactive, inquiring if all was well.

The subtler lesson was the effect our personal scores had on us as individuals. At a certain point in the challenge, it was clear that nobody was going to catch the 60 and 73 year old men that were leading. Yet, everyone stayed engaged. I circled back with some of the participants and found they all hated to see that dreaded down arrow show up next to their score. It meant they weren't active and their score was trending down. They were competitive with themselves, and found they were striving to be the best version they could be.

I probably should have seen that coming. I have a buddy I golf with all the time and I realized, sure we enjoy bragging rights but more than anything else, we just want to beat the golf course. On any given day, one of us might play out of his mind and take home the trophy but over time, if we focused on becoming the best we could be, we were satisfied.

But with so much that is broken in the world of health care, where do you start? Let's start by looking at how our health care system is oriented today . . .

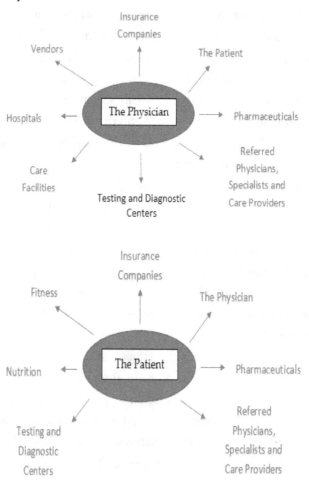

And let's next look at how the health care system will look like in the future:

Our current system is built like a solar system where the physician is at the center, with insurance companies occupying a close second place in the hierarchy. What is wrong with that? Consumer health care should occupy the central position. Health care payers, providers, insurance companies and the rest should revolve around the consumer.

Judging from the comments offered by many of the experts who contributed to this book, this is already happening - larger structures are reforming around the consumer. In a sense, that is what this book is about - how we need to reorient health care's value chain. That sounds simple in concept, but the change from the old to the new will require nothing short of a massive transformation.

A ROADMAP FOR CHANGE

Let's first take a look at where we are now. I find it helpful to conceptualize the changes of digital health disruption in terms of three phases.

DIGITAL HEALTH 1.0

This is where we dispelled the popular myth that people won't engage or share. Thousands of health care apps appeared overnight, and marketplaces flourished. The first wave of health tech innovation designed from the ground up to treat health care like a shopping experience came from outside of the industry!

"We just can't get people to engage" has become yesterday's battle cry. If anything, there seems to be a growing appetite among consumers to become more involved in their health care. Health care consumers are donning Fitbits, buying bathroom scales that keep ongoing data about their daily weigh-ins, and taking other steps to gather and monitor their data.

People are more engaged in their own health care and wellbeing. This explosion of digital health apps dispels the popular myth that people won't engage or be held accountable.

Surprise! They will engage . . . they already are.

DIGITAL HEALTH 2.0

Welcome to the quantified self movement - and a device for everything. Big Data dominates top-of-mind thinking as analytics platforms for benchmarking and predicting better outcomes emerge. The smartest men in the room are looking backwards and explaining everything that just happened, but believe data isn't actionable--it's for analyzing.

DIGITAL HEALTH 3.0

Welcome to connected health. We are on the cusp of this now. Consumers are engaged and collecting data that is being fed to analytics platforms. The focus shifts from understanding what just happened to how to manage better outcomes.

This will be the tipping point when the consumer will move into the center of the health care universe. It will represent the biggest opportunity that we have ever seen to revolutionize health care and enhance patient wellbeing.

This will be where we, as medical consumers, make our data actionable by connecting it to the stakeholders who help keep us healthy. This future health care will center on coordinated care teams who will use the data captured by digital health assets to keep consumers well, not treat them after they become ill.

You see, in the past, the term "health care" has been a misnomer – it has really been "sick care." But because of the coming consumer-driven digital health care revolution, this is now changing.

MARKERS ALONG THE WAY

The Affordable Care Act is already transitioning us from a "fee for service" model (a patient pays for a service) toward a value-based care model. But more steps need to be taken.

SHIFTING THE PARADIGM FROM
CURING DISEASES TO KEEPING PEOPLE HEALTHY

Like many daunting challenges, fixing our health care system is a complex problem with a pretty simple solution:

We need to reorient our way of conceptualizing health care.

We should stop expecting our system to cure people after they have become sick and start thinking of it as a way to prevent people from becoming ill.

Presently, care providers take very few steps to help patients who are not standing right in front of them. At best, they give those patients tracking tools they can use to know better when something is going wrong, and when it is time to call to make an office appointment. It is still a "give me a call when you are broken" mentality.

That will change. There will be disruption, a very positive disruption. To get to the place where we need to be, we need to question many of the operative assumptions that have taken hold in our health care universe.

Providing easier and faster access to physicians, while a huge step forward, will not move the needle enough. And sad to say, simply revamping insurance, while essential, will not go far enough either. Even the most miraculous technologies like the sharing of medical records across computerized telemedicine platforms, although essential, will be only a component of tomorrow's miraculous medical world, but not its salvation.

WHAT ARE THE DANGERS OF
NOT FIXING OUR BROKEN SYSTEM?

The greatest danger could be complacency. Or perhaps more accurately, a kind of torpor in which we become tone deaf to chronic conditions – something that can happen if we allow our brains to become so buried in the idea that chronic conditions and diseases of the past are still incurable today. It is time to accept the simple fact that many of the problems that face health care today are more in our control than we realize.

We could have lost our way because so much has changed, so quickly. We are absorbing so much new information about health and medicine. Our way of framing problems often lags and fails to absorb the new realities that lie right in front of us.

We've just started living in a new millennium, and we sometimes fail to notice that the problems we are dying from today are not the same illnesses we were dying from a century ago. Back then, we were dying from diseases like tuberculosis, the flu or polio, and they decimated us.

Many of the most widespread ailments in those days were diseases that people contracted when they left their homes. A patient went out into the world and got a viral infection, and there wasn't too much he or she could do to keep that from happening, or to cure it once he or she did. Today, we are becoming ill from a smaller number of diseases that we contract as a result of our daily lifestyle choices. Although there are still cancer, AIDS and a handful of other diseases, we are mostly dying from things that we do to ourselves.

We gain weight, eat the wrong foods, fail to exercise and fall into bad habits that result in adult-onset, type 2 diabetes, heart disease, circulatory problems, and a number of other issues. The fact is, we can control these variables, and we are living in a time when we can prevent a larger percentage of diseases by taking preventative steps that are proactive, proven, and very, very doable.

In the past, people couldn't prevent today's most common ailments because they didn't know how. Today, the health care industry can show consumers what to do. And consumer appetite to do the right things is already there.

When you layer data collection and monitoring on top of those trends, you can see that the stage has been set for Digital Health 3.0.

CASE IN POINT: A PATIENT WITH TYPE 2 DIABETES

Chances are you know someone who has developed type 2, adult-onset diabetes. This disease has become so prevalent today, it is a certainty that you do.

Let's take a moment to consider how that person you know is approaching different options for treating his or her condition.

Under the old "wait until a patient becomes ill and then treat them" paradigm of years past, few care providers thought about preventing this disease before it emerged. After the patient developed troubling symptoms that demanded a diagnosis – fatigue, sleep apnea, circulatory problems, sexual dysfunction, vision problems or worse – a range of treatments was suddenly started. The patient was put on a diet, given a list of starchy and sweet foods to avoid, given prescriptions, perhaps assigned a nutritionist, and started on a program of exercise. And of course, he or she got a stack of pamphlets and books to read about how to manage the condition.

That might all be to the good, but there is a better, preventative way to administer health care. If that same patient had been monitored by a physician and other care providers who were watching for weight gain, fatigue, lack of exercise, poor eating habits and other factors that can reliably predict the onset of type 2 diabetes, that patient might never have become ill at all.

This is more than theoretical, as you will learn when you read the contribution that Sean Duffy of Omada Health made to this book. Omada is at the forefront of helping pre-diabetic people never get the full-blown disease. And they are using digital tools to do it.

WHY DID THE OLD WAY DEVELOP?

I don't mean to be cynical, but couldn't the underlying problem be that in the past, physicians, hospitals, insurance companies and other entities in the health care universe only profited when patients became ill?

I am happy to observe that that paradigm is changing. A growing number of health care providers who focus on preventative medicine are finding their place in the health care landscape. Providers like nutritionists and exercise therapists are growing in number. But the core principle, even today, is that most providers only make money when people get sick. And that must change.

WHY TECHNOLOGY ALONE IS NOT ENOUGH

We live in a time when diagnostic testing has reached new levels of effectiveness. A growing number of ailments – aneurisms, blocked arteries, colon cancer and more – can now be discovered before they cause major damage to patients.

Similarly, medical informatics and telemedicine have brought a new level of effectiveness to medical care. If you are admitted with a stroke to a small, remote hospital, for example, your case can be reviewed by specialists who work far away. Or if you suddenly fall ill while you are on vacation far from home and are admitted to a hospital, chances are that your records will be immediately available to the physicians who are treating you there.

There is a very real "wow" factor when we see modern technologies like those at work. They make us think that the quality of our medical care is already miraculous. And in many ways, it is.

But I would like to submit my opinion that technologies – even the greatest and most powerful among them – do not have the power to revolutionize today's dominant approach to health care or bring consumers into the epicenter of that solar system which we discussed at the start of this chapter.

Technology, even powerful technology, will not move consumers to that central place. We need a fundamental reordering of priorities and a complete shakeup.

CASE STUDY: A MODERN PROVIDER OF ADVANCED SURGICAL PROCEDURES

I do not want to name the service provider I am about to describe. Let me just say that it is a chain of surgery centers that offers an advanced, immediate surgical solution to a medical problem that is common among tens of thousands of medical consumers who live in all areas of our country and the world.

This provider offers a vastly improved, minimally invasive surgical option to patients, many of whom travel long distances to undergo this company's advanced surgery, which promises dramatically shorter recovery times and a quicker return to work and normal living.

That sounds like a major step forward, doesn't it? And for many patients, it is. This company aimed to make its services available to a dramatically larger number of patients who pay very high fees for the surgery. They increased the number of patients they treat, and optimized the procedure itself. However, they did not necessarily improve post-procedure results for patients or educate them on how to avoid injury in the first place. In my opinion, their motivation has been to build a company that has the potential to earn a staggering amount of money, not to transform the way we care for people with back problems.

Now, I am not saying that people are not being helped by their surgeries. In many ways, what the company offers is a win/win for both the company and the patients. It is just that this provider is not doing much to teach people how to avoid having their medical problems in the first place. It is not saying, "here is what you need to do today to avoid becoming one of our patients tomorrow."

Today, providers specialize to better maximize profits and streamline operations. In the future, health care will be all about detecting, anticipating, and preventing by using advanced digital techniques.

WHAT CAUSED OUR SYSTEM TO VEER OFF COURSE?

The desire to help and cure patients is the reason many people choose to become physicians and enter other curative professions.

But where do those idealistic people end up? What roles are thrust upon them when they begin delivering health care? Regulation, the need to be a profitable network provider within their districts, the necessity of qualifying for insurance repayments and other factors constrain them to take on different, less idealistically motivated roles.

Lots of tensions and conflicts enter the picture. There are tensions between payers and care providers, between patients and insurance companies, and between other entities within the mosaic of health care. Medical care becomes a fungible good that providers want to sell to consumers. Physicians and other care providers are part of networks, and therefore are limited – or at least, directed – to recommend in-network care providers to their patients. Consumers, who are called "patients" (and who are relegated to that role), are within insurer-mandated networks and, therefore, are part of a captive audience who can pick and choose from only a limited number of providers.

I think that is all changing. But I also believe that it will take a highly disruptive process to move to a system that puts the consumer at the epicenter of it all. This process is not going to happen overnight, but many of the thought leaders who have made contributions to this book are driving that change.

EXPERT OPINION: DAVID C. EDELMAN, CHIEF MARKETING OFFICER, AETNA

In the minds of patients and consumers, insurance companies once seemed to occupy a secondary position in healthcare. Consumers visited physicians, hospitals, testing centers and other care-providers and thought of them as the enablers of their care.

If you asked those customers what they thought of their insurance companies, most would have said, "They are there to process my medical claims." The notion that an insurance company cared about their wellbeing or wanted to keep them healthy was simply not on anyone's radar.

But while customers were looking the other way, a seismic change has taken place in the role that insurers are playing in their wellness journey. And even bigger changes are coming.

When Aetna hired David C. Edelman as Chief Marketing Officer in October 2016, the company was looking for someone to refocus Aetna's marketing on the individual customer.

"I was hired to transform Aetna's marketing organization, creating meaningful consumer experiences and new initiatives to unlock growth," Dave told Kevin Pereau recently." We are only at the beginning."

But what a beginning. Dave and his team are succeeding by applying both proven ways of marketing insurance and thinking that is new to the insurance industry. Perhaps that is not surprising, given his background. Before joining Aetna, Dave co-led the Global Digital Marketing and Sales Practice at McKinsey, where his cross-functional team focused on digital marketing, analytics, and process redesign. Prior to McKinsey, Dave was executive vice president for the Strategy and Analytics practice at Digitas. Earlier still, he helped to build the Boston Consulting Group's "Segment-of-One Marketing" and "E-commerce" practices.

I am delighted that Dave agreed to discuss with me some insights on what is going on at Aetna, and to share some of his keen insights on the changing world of healthcare too. Let's hear what he has to say.

HOW AETNA IS REFOCUSING ITS MARKETING

Marketers, and digital marketers in particular, have developed and perfected tools to optimize and measure the effectiveness of their marketing. The strategies are to develop marketing programs, test their effectiveness, and then optimize them to improve results.

In Dave's words, those skills are "fundamental to our day-to-day work. We continue to work closely with our business partners to raise the bar on the traditional marketing to support our sales force.

Yet things are changing. At Aetna, Dave and his associates are shifting the focus toward finding the right creative ways of connecting with individual consumers - not to just sell to them, but to support the goal of helping them take better care of themselves.

Dave explains that there are two sides to how Aetna is reshaping its marketing. One is engagement, which Dave defines as "Using marketing techniques to get people involved." The other side is experience, which Dave summarizes as, "Making people's experience with us something that's distinctive, branded, and worthy of sustaining over time."

The success of that process will mean getting people to become more engaged in managing their care, in becoming more aware of the decisions that they can make and the tools that they can use to take greater control over their own healthcare.

"We need to bring marketing discipline into the actual care management process," Dave summarizes. On the surface of things, that might sound straightforward. But when I take a moment to think about it, I realize that Dave is talking about a change that is profound.

REINVENTING THE AETNA BRAND

Dave and his colleagues have launched a new brand positioning that is simpler, more direct, more active, and livelier.

"We're not just about sick care; we want to be about healthcare and helping people realize the joy of achieving their health ambitions, whether that's running a marathon or attending a family wedding" Dave explains. "So we dedicate every ounce of our energy to helping people reach their health goals, big and small, at every stage of life."

That new emphasis is reflected in the new tagline that is central in the marketing programs that Dave has begun at Aetna . . .

"You don't join us. We join you."

To make that brand promise come authentically alive, it is critical to understand and support individual consumers' individual health ambitions. It is a process that will require rethinking the entire member experience.

"In some ways, you might say we are not the marketing department, but `the office of the consumer,'" Dave explains, "even though that is not our title."

TURNING THE INSURANCE INDUSTRY ON ITS EAR

In years past, helping insured customers become healthier was something that just kind of happened because they had insurance coverage. Now that is changing to a new paradigm in which the individual customer's health journey will move to the very center of what insurers do.

"There is a growing realization that for health insurers to survive, they cannot just pay claims," Dave states. "The old model was insurers would get a bucket of money from an employee or an employer or from the government, and they would spend some of that to play claims, and whatever they could keep was profit."

Predictably, the insurer's ability to be profitable depended on spending as little as possible to cover claims. And other unfriendly practices took hold, such as making customers expend lots of effort to find in-network care providers, to avoid certain costly procedures – and worse.

"But there should be more to our business than just being parsimonious about paying claims, which often didn't help the consumer very much" Dave states, noting that today's consumers are increasingly interested in assuming more responsibility to manage their conditions, to manage costs, and to make very informed decisions about their care.

Can this increased customer involvement create a more efficient, more profitable way of providing insurance? Dave believes that it will. "There were, in fact, enormous costs involved in doing business in the old way," he states, "because people were just bouncing around like bumper cars or balls in a pinball machine, finding different care providers in the healthcare system. There has been no one place to coordinate that care within the clinical complex, and beyond the clinical complex."

When an insurer can understand informed consumers' concerns and partner with them to make the best decisions regarding the care of chronic and other conditions, both better care and reduced costs can result. Both those goals can be reached by shifting the emphasis to helping customers stay healthy instead of paying claims to help fix them when they become ill.

THE ROLE OF DATA

In the past, Dave points out, insurers collected data about customers' care not only to increase profits, but to be responsive to employers who were providing coverage, or to the government, which might have been paying Medicare expenses.

"Insurance companies' systems were not originally crafted to gather and use individual data to the benefit of the consumer," Dave explains. "They were mostly gathering information on instances of ER use, on this, on that . . . and just basically reporting back, and using data to make some decisions on where to focus more on utilization reviews, things like that."

Was useful data being gathered? Yes, but the purpose of the gathering was not to help individual consumers proactively deal with their health concerns. Now that is changing. Aetna is developing systems that are acting as smart, cognitive databases that alert customers to ways to improve the quality of their care.

Dave summarizes, "We are re-architecting our systems in ways that provide customers with constant, optimized messaging that leads them to engage in smarter ways to help themselves. And we have actually gotten quite far in that process."

DATA FROM DAY ONE

As soon as a new Aetna customer comes on board, Aetna asks "just a few questions" to make sure that the company understands their conditions, concerns, the health options they are utilizing – and can suggest appropriate care options they might not be utilizing.

"And then from that," Dave says, "we can decide about what level of support they will need, and what kind of support. And then within that support, we're going to track certain things to see if somebody is on track. If somebody who is a diabetic has not had their A1C checked in a while, we want to immediately get on that and trigger them to go into a clinic someplace and get their A1C checked. We want to translate this data into active triggers that we can use to help people."

There are wider ways to use that data too. If Aetna suddenly sees a surge of flu claims in a region, for example, "we are going to work across the company to encourage members in that area to get flu shots," Dave says.

The objective, which Aetna is already meeting, is to use data as something active that can be used to help keep members healthy.

Do members sometimes worry about sharing their data? "There are privacy concerns." Dave says, "and they are valid. But our intent is to focus on achieving positive health outcomes for the member. That's the standard that we are using. They choose it because of care, because of health."

BRANDING TO BUILD LOYALTY

Aetna foresees that those points of contact will provide opportunities to build durable loyalty and affinity for its brand.

For example, Dave believes that as they move forward in their lives and their careers, younger consumers will have more opportunities to choose their insurers. "More employers are offering not just one, but many healthcare options, "he states, "and so people have choices. We want to be the brand that is preferred, and for reasons that go beyond the basics about cost and network coverage. We want it to be about the experience and what we do to help people's care."

In part, opportunities will arise from what insurers in the past have seen as concerns. In the increasingly complex world of insurance, individuals will have more responsibility to decide how they will pay for care. "There will be more interaction because consumers are getting more bills," Dave explains. "They need more help, so they can make smarter decisions for themselves. On one level, they need help to stay in network, to manage costs, and to find their way to the right care. The result is that they are expecting much more from their insurance companies because they, as individuals, have more responsibility and need more help."

But underlying those complications and concerns, Dave sees an opportunity to make consumers feel more intimately connected with

Aetna, and more loyal. That connection is the key to Aetna's evolving future.

BREAKING DOWN THE CLINICAL/INDUSTRIAL COMPLEX

"I think that one important trend is the breakdown of what I call the clinical/industrial complex," Dave comments, "It is very easy for people to get lured into big, high-tech medical complexes when what they really might need is just a really good massage therapist, or other forms of alternative support. Maybe they need a nutritionist at a Safeway store who can guide them toward the right foods when they are in the store. I think we're going to see, and are already building at a community level, extended networks that go beyond the clinical complex to include other forms of support that help people in their care – and in their health. We'll also see more of those networks formally becoming parts of networks in a more rigorous way. It's just starting. I see us doing it. I think others will too.

"It will go beyond just saying, `You'll get a few dollars off a gym membership as an Aetna member.' It's going to be more explicit interconnection to new, more extensive, alternative forms of health support."

IMPORTANT TRENDS

Exercise clubs, pharmacies, exercise clubs, nutritionists and more will all be brought into that matrix. And in their marketing, insurance companies will move beyond talking about coverage and fees paid and tell more of the stories about how they helped individual people in their health journeys.

But perhaps the biggest trend will be that the emphasis in insurance will move from paying bills to partnering fully and actively with consumers to build wellness . . . to build health.

"Everything in the past was about transactions and driving down the cost of those transactions," Dave said as our talk came to a close.

"I think you're going to see a re-centering of the entire business around driving outcomes which can be summed up simply as, 'health.'"

EXPERT OPINION: CHRISTINE PAIGE, PHD, SENIOR VICE PRESIDENT MARKETING & DIGITAL SERVICES, KAISER PERMANENTE

If you ask people what Kaiser Permanente is, chances are they will say, "I've heard it's one of the very best health insurance providers in the United States."

That is correct, because Kaiser is deeply committed to delivering top-notch plans and service to its members. But Kaiser is much more than that. It is, in fact, a unique nonprofit organization that offers not just health insurance, but a network of top-quality clinics and hospitals that are staffed by Kaiser-affiliated physicians and caregivers. Kaiser, which was founded in 1945 by industrialist Henry J. Kaiser, has grown to be the largest managed-care provider in the United States. Kaiser currently has more than 12 million health plan members, more than 22,000 physicians, and more than 55,000 nurses who provide services in more than 40 medical centers and about 700 Kaiser medical facilities. Kaiser Permanente is, in fact, a completely unique organization that brings unique excellence to everything it provides.

Christine oversees advertising and brand management, marketing content and communications, direct and digital marketing, consumer and customer insights, customer value reporting, and proposal development. Paige is also responsible for leading Kaiser Permanente's consumer strategy, focused on building excellent consumer experiences. She has also led the development of e-commerce capabilities and web and mobile clinical features for Kaiser Permanente members.

In these capacities, Christine has been at the forefront in establishing Kaiser Permanente as an international leader in the use of personal health records and informatics to enhance and expand affordable, high-quality care and service. She was instrumental in developing the landmark "Thrive" brand advertising campaign, now in its fifteenth year. The award-winning campaign embodies Kaiser Permanente's commitment to total health, through an integrated health care delivery system, focusing on preventative care and the empowerment of members to maximize their total health — mind, body and spirit.

After joining Kaiser Permanente in 1988, Christine has served in a variety of roles, and has led the organization's marketing functions nationally since 1999. She is a frequent speaker on the topics of technology, consumer engagement, and marketing.

Prior to joining Kaiser Permanente, Christine was a political science professor at the University of Notre Dame and Southern Methodist University, specializing in comparative political development, and Italian politics.

Christine received a Bachelor of Arts degree from the University of Southern California and master's and doctoral degrees in political science from the University of California, Berkeley.

CHRISTINE PAIGE ON KAISER'S INCENTIVES AND ORIENTATION

We stay focused on the health outcomes and satisfaction of our members, rather than on factors like our transaction volume. And to me, that orientation is the key to everything. It keeps our people focused on what matters, rather than on simply what is doable. And it makes it easier to focus on the patient care experience, because considerations like transactional activity and financial data fade completely into the background and become invisible to the people who are in the system.

That orientation was easier for us, because we were an early believer in maintaining electronic records for our patients. Because we were an early mover, it was easier for us to make the patient experience central to what we do. We understood that the benefits of investing in patient data would accrue back to us.

And globally, we got that right as an organization. Whereas for the rest of the industry, the barriers to early investment were often about who was going to put up the dollars to invest in technology, and what the benefits would be. As a result, for those companies there were all kinds of aspects of the care experience that never became integrated.

ON HOW TECHNOLOGY RELATES TO DIFFERENT POPULATIONS

I think there are several different dimensions to thinking about how technology relates to populations and their health management. With the right technology enabled, screening for preventative health became a bedrock of our operations. It is how we are able to catch cancer early, prevent heart attacks, and make very significant improvements in people's lives.

It depends on making sure patients are screened, making sure we know what their risk factors are, so we can intervene earlier with them. We have a digital backbone to our operations.

We have numerous stories of people coming in to get their eyes checked, for example, but the receptionist reviews their records and says, "Did you know you are really overdue for your mammogram?" And the patient will say, "yes I know, I will get to it." But and the receptionist says "Well, can we make that appointment, I'll pull it up right now . . ."

We save lives that way. That's the importance of having disciplined execution tied to data and preventive care. We can cover an entire population in a way in which digital tools help the provider, and help everyone in the system execute on that. So that's a very important point.

ON PATIENT ENGAGEMENT

In *The Digital Health Revolution*, you point out that in years past, it was difficult to engage people to become involved in managing their data and their health. Yet there are some segments of the population that are active in self-management. We call them health-seekers or information-seekers. They will read everything, and they will research everything, right? And you know, that is 20% maybe 30% of the population. But the rest of folks also want their care experience to be simple and easy.

In the early days when we were simply offering information online, we learned that it was difficult to engage with people. Then we began to offer transactions, time-saving opportunities. People could manage their prescriptions online, make appointments and message with their physicians. As a result, we have grown to the point where 73% of our members are registered online with us, which leads the industry by a very big margin.

MEASURING THE RESULTS OF ONLINE ENGAGEMENT

We are a very evidence-based, research-oriented organization. One thing we investigated early on was the question of whether patients who are engaged online actually become healthier.

In one study, we investigated whether our members who are online have diabetes have better blood sugar control than members do who are not online. When we investigated, our data showed very clearly that patients who are engaged online have much better blood sugar control than patients who are not online.

Why would that be? It is because they are more likely to make appointments with their physicians, more likely to keep up with their medications, and more likely to engage in a whole range of healthier behaviors. So we certainly see clinical benefits as well as higher satisfaction levels for people who are engaged online.

Another thing our research has found is a business benefit, because the people who are engaged online are more loyal to Kaiser. We have higher retention among those members then we do with people who don't engage.

ON DEVELOPING TECHNOLOGY
IN-HOUSE VS. OUTSOURCING

One example of sharing technology is that we use Epic 4's services for electronic medical record-keeping. That company provides the underlying functionality that is part of what's on our member portal. Our digital health monitoring modules come from Johnson & Johnson. Some of our health content comes from Healthwise, and we have a long-running relationship with American Specialty Health. We have worked cooperatively with Omada Health to provide some of their diabetes options for our members.

How do we decide what to bring in from the outside and what to develop ourselves? We go through a technology evaluation process that is essentially like our pharmaceutical and therapeutic evaluation process.

ON HOW POTENTIAL PARTNERS
COME TO KAISER'S ATTENTION

When companies contact us with an interesting technology that might be good for us, we engage in what is essentially a process of gap analysis. In other words, does what they are offering help us better meet a specific need of our members?

We know, for example, that some of our members could use help with transportation they need to get to their appointments. So when someone comes to us with an Uber-like app that could potentially be incorporated in plans, we might consider that from the perspective of whether it improves our capabilities, or introduces a new option that would be meaningful to our membership.

ON SOCIAL DETERMINANTS

Social determinants have been a major area of focus for us for seven or eight years now. Because we are a not-for-profit organization, we have a mission around community health and a long history of making investments in programs that promote health, through nutrition, community health clinics, a wide range of things.

In the last few years, we have focused on a wider range of activities like making sure that our lower income members are signing up for programs that will benefit them. We have just announced a $200 Million investment in addressing homelessness through impact investing. So it is fair to say that we want to improve health at both the individual and the community levels.

I am sure you have seen studies that show that people who live in specific Zip Codes often suffer from the same conditions. We look at patterns like that and see that addiction or other factors within a specific area could be improved with the right kind of initiatives from us.

For more than a decade, we have seen opportunities that we have been able to take advantage of because we have been part of the digital world.

ON RESULTS IN FIGHTING SMOKING

Our members have among the lowest smoking rates of any health organization.

A big reason is that it is virtually impossible for members to get through any interaction with us without being asked whether they are still smoking. It is a question we focus on as part of our basic intake procedures, but it something we ask in virtually all points of patient contact. We ask about alcohol use. We ask about exercise, which we call a "vital sign."

And incidentally, thanks to our data, we have among the lowest rates of opioid abuse among our members. Our patient records are critical there. We can look at patients' prescription records to evaluate whether patients have been overusing painkillers. And our data assures that our members cannot get their prescriptions filled at multiple pharmacies, or visit several different physicians to get refills.

Many people and organizations have data that is really robust. But the issue is really how you use it, right? And how you turn it toward solving problems.

ON HOW KAISER WILL BE USING DATA IN FIVE OR 10 YEARS

There are thousands of apps out there that are meant to do the right kind of tracking, issue reminders and do more, and they will continue to grow in number and scope. We've done some work with telephonic health coaching around nutrition, stress, and so forth. But for us, the major focus with data has always been the individual health of our members. And I think we are going to continue to innovate around that. I envision that even more powerful opportunities will emerge from that focus.

What we are working on around digital innovations is the question of how we can continue to make the user's experience just that much better. We already have an organization where our members, from the moment they open their phones, can make a face-to-face or a telephonic appointment, or can get real-time advice.

We're moving into a world that is much more preference-sensitive, where our members can shape their own experience, and where there are more modalities to choose from. Overall, the experiences that our members can have, either in a physical environment or on their phones, are merging.

When you arrive at a physician's office, a clinic, or a hospital, you will be able to register on your phone instead of at a desk. You will use your phone to connect to the pharmacy, to pay, and to take advantage of all the benefits of connectivity, period. And that's the sort of future state we are working on now,. We have high digital engagement rates now and are really, verifiably improving the health of our members.

But in the future, I believe the experience of healthcare we are moving toward will be one that is highly personalized and utterly convenient.

EXPERT OPINION: RAJEEV RONANKI, SVP AND CHIEF DIGITAL OFFICER, ANTHEM INC.

Rajeev Ronanki is Senior Vice President and Chief Digital Officer of Anthem Inc., a Blue Cross Blue Shield company that is dedicated to providing consumers with affordable, high-quality health care. Anthem currently provides more than 74 million people with health care, about 40 million of whom are enrolled through family health plans administered through Anthem's affiliated companies.

Rajeev has spent more than two decades in health care. Prior to joining Anthem in 2018, he spent a decade at Deloitte Consulting, where he led Deloitte's Cognitive Advantage (Markets and Technology) and Health Care Innovative practices. While there, he focused on implementing solutions for personalized consumer engagement, intelligent automation, and predictive analytics.

Kevin Pereau: I'm wondering what Anthem has been up to with developing new solutions for connecting with health care consumers. We hear a lot of buzzwords in health care today, words like seamless, consumer this, consumer that. If you will drill down into it with me, what do those terms mean? And what do they really mean for individuals who are Anthem clients, either individually or through employer and other plans?

Rajeev Ronanki: I'll use the words technology, analytics, artificial intelligence and digital—probably all part of a spectrum. We're going to be using a combination of those technologies for our stakeholders. But if you look at that packet of technologies, I think I'll use those terms generically to be able to simplify our conversation.

The heart of it? At the end of the day, I think Anthem is a data, analytics and AI company. It just so happens that insurance is the business model that presents our value proposition to the market.

Let's think about Google for a minute. Search is the core technology that Google has. But at the bottom of it all, Google is offering advertising and marketing and a matching set of services that are part of the overall value to consumers. So we think of insurance in sort of the same way, and we offer insurance products.

Apart from the individual market and Medicare and Medicaid, for the most part insurance is still largely a B2B business. And flowing from that, our focus and intent are to serve our employer groups. That's who we're contracting with. And the members are somewhat incidental to that, ultimately the consumers of the services we are creating in conjunction with our employer groups.

But the consumer experience needs to be much more than that, because no one is happy living in that model. Consumers feel like they can't really navigate the complex ecosystem between providers and life sciences companies, or paying for insurance, or a number of other processes. There's no transparency for consumers about what the real costs are and who is really paying.

One significant change is that consumers are starting to take a more active interest and asking questions like, "Well, what am I paying for? What am I getting in return for it? And why is my experience in health care so poor, compared to, let's say, Uber and Amazon or

Apple, or other consumer-oriented companies that provide world-class service, at seemingly no cost to the consumer?"

And therefore to survive and thrive as a business, we have to orchestrate the entire suite of services around the consumer, even if there is an underlying B2B model. It has to be consumer first. And then the solutions come from there, instead of putting consumers somewhere toward the end of that spectrum.

Does that make sense as a frame to explain why we are orienting technology toward the consumer?

Kevin Pereau: It does indeed, and it gives us a great framework to discuss. We're probably eight years into the investment explosion into health care. But when you put your thinking cap on, how do you decide whether to partner with a company like VIM, or with an Omada? And when do you decide, "Instead of working with one of those companies, we are going to build this functionality ourselves?"

Rajeev Ronanki: That's a fair question. If you look at the Apples and the Amazons and the Googles, you see that like them, we are not ultimately building everything in-house. We are in a broader community and a broader ecosystem that connects a common data fabric. How each of those companies automates is different. But there are some very common patterns you can apply to a platform business. And then the economics that follow from that can be compelling.

So think of Anthem as being a data platform. Which means that we will certainly build things that are absolutely essential to the member experience. Anthem will obviously address other entities and stakeholders as well, but at the moment, the focus is on the consumer. We want to create an optimal experience for members by first asking, "Well, what would that experience consist of?" And then to ask, "Do we have the data to do it justice?"

If the answer to questions like that is yes, we will build it. As an example, let's look at scheduling, a core consumer activity. Improving it is going to require us to cooperate with all the other providers and their practice management systems and their scheduling systems. The question is, how can we simplify the task of making an appointment? Making an appointment online is easy. But it might not be something that we would want to build a system for ourselves. So we may say, yes, let's partner with a certain company, rather than building a scheduling system ourselves.

But for other things like core engagement, we are asking, "How do I continually know the consumers better and start to create a more transparent and personalized experience for them?" And we might find that is something we'd want to build.

Think about what Amazon has, that kind of shopping experience. You log into Amazon, you discover something you want to buy, and you buy it. That's pure Amazon IT. But for shipping and for other things, Amazon is working with a number of partners. What we are doing is very similar to that.

Kevin Pereau: That makes total sense. In my book, I've been writing about how we are hurtling toward a world of connected health and making data actionable. It sounds like Anthem is moving in that direction as well.

Rajeev Ronanki: Oh, absolutely. I think data is our primary asset. And we're using it in a way that is very responsible. I don't know if this has come up in any of your other interviews, but we are taking steps to make sure that we promise data privacy.

Responsible use of our algorithms is one of our core values and orientations. We eliminate biases in our algorithms—how we make decisions with data. It's very transparent. As the systems that are built on them become more and more sophisticated, we are able to trace back and say, "Well, why is it that we're making these recommendations?" And it is important to have that information available so the consumers, the members, the providers, all can peer into that and decide if that's what they want to do. It's aligned to the collective interest of all parties, not just Anthem's.

Kevin: Pereau: I'm an ecosystem guy. I was a Cambridge Technology Partner, where we were all about ecosystem and leveraging the greater sum of our parts. But how do you get all the parties in the ecosystem to share your mission and participate in the value creation?

Rajeev Ronanki: That's the curation part. In our case, we want to be very thoughtful about who we choose as partners. And we go through a significant amount of alignment from the perspectives of privacy, data rights, usage and more, to make sure values are in synch. And then, we obviously build our agreements with the goal in mind to make it financially viable, so all parties benefit from the relationship.

But let me stress, that doesn't mean that we are not going to work with third parties. More and more, the companies we are working with are aligned around making sure that the data relationship resides with the consumers. We want to be able to very transparently and very openly communicate how we use their data. As long as that common Bill of Rights, so to speak, is in alignment with our ecosystem, we will absolutely work with them.

We are not going to work with companies that want to take that data and drop it into third party sources. That might be part of their business model, but in cases where we can't control where the data ends up or how it gets used, we will not become involved. We need to assure our members that there will be no malicious use of their data by some party downstream. So we're building in a lot of data governance, to be sure the partners that are coming into our system share the same goals.

Kevin Pereau: What is Anthem doing, and what is its activity on the investment and venture capital fronts?

Rajeev Ronanki: We do have an internal venture capital department. We are not scaled to be a traditional VC kind of company. But we are able to say, "We like these two or three bets here. So let's go ahead and make an investment in them."

We'll continue to expand that. Because we see that as a critical shift. Again, building all the resources ourselves is unrealistic.

Kevin Pereau: I'm sensing that the industry can't remain focused on the low-hanging fruit of finding solutions for diabetes and hypertension and similar ailments that affect broad segments of the population. It seems the industry is turning a corner on recognizing that there's a whole untapped market segment they're beginning to take a look at. Are there any populations that you particularly want to focus on?

Rajeev Ronanki: I'm not sure if it's a segment of the population that we're focusing on, as much as focusing on whole health, from physical to emotional wellbeing. And financial wellbeing.

We believe we have to deal with the whole person. Much of that will involve how we deal with incentives and rewards, get to the biometrics where appropriate, and create a feedback loop for the physical wellbeing and emotional wellbeing, which is getting more traction these days.

Financial wellbeing is another sort of thing that is closely tied into the social determinant. We had a case recently where a member was unfortunately diagnosed with cancer. That's one big life event. But at the same time, the consumer is stuck with a huge medical bill as well. That starts to impact the family life. It is time to start thinking in new ways about kids and all. There's an entire spectrum of issues that we need to piece together in order to serve the member well. That's where we're going.

The population of health care consumers is very diverse, and we're not excluding anyone. We've got people who are gig economy workers, people who are aging into Medicare, people with chronic conditions, and others. But you've got to look at overall health to serve any segment well.

Kevin Pereau: Do you think we're turning a corner so that larger healthcare insurers and other businesses will be better able to serve people from a variety of populations who have different health concerns?

Rajeev Ronanki: I certainly hope so. We're still in the early stages. I'd say the major players in the ecosystem are all recognizing that is the end goal, just to get there.

Kevin Pereau: And where is Anthem heading next? What are the priorities?

Rajeev Ronanki: If you look at our net assets, the data assets, we've done a lot of work in the provider network optimization space. We've got a huge library of algorithms and we're doubling down on creating even more. And there are some really innovative ways we may be able to match providers to their members, much like how Netflix matches movies to consumers. We want to get to that next generation of being able to match providers and consumers, focusing on cost and quality. That's going to be a core piece of what we're going to do.

Member engagement, as we're calling it, is going to have a lot of personalization built into it. And what I mean by that is, we take a longitudinal view of our members' health data that includes social determinants, that includes claims obviously, and that includes a lot of clinical data. And increasingly, a lot of biometric data from devices too. So if you put that all together, there's a kind of rich view that gives us all kinds of insights.

That is going to get built into a very tailored data set that leads to proactive outreach that enables us to say to consumers, "Here's what we think you ought to do next." If in fact members want to be communicated to in that fashion. Not all people do. For those people, there will simply be a kind of digital breadcrumb that invites them to the next thing they ought to be able to do as they manage their care. And built into the system is what they can choose to do, and what happens afterwards—medications, specialists they need to see. Our vision is to be able to serve as a guide to seamlessly make all of that happen.

Kevin Pereau: That's really exciting. It's great to see you evolving a framework and a strategy around that. As you think about the vetting process and all that you just laid out, what's the best way for technology partners to approach Anthem? How do they land on your radar?

Rajeev Ronanki: We have the capability to ingest a lot of data from VC firms, from our network, and elsewhere. As you would imagine, literally hundreds of ideas and solutions come our way. We organize that information into what you could call a taxonomy of solutions. They could be anything from care coordination to virtual care, to telehealth, to condition-specific apps like VIM. We sort them into one or more of those categories. We've got a sensing algorithm that enables us to go through opportunities that other companies offer and say, "This looks like it fits a need that is open, should we look more into it?" And then we do. And if we find it interesting, we might reach out and have a discussion with the company.

An example of that is a recent filing that we're doing with a company that wants to do personalized clinical trials. So clinical trials are one of the categories that we really didn't have any near-term needs for, but this company approached us to say, "We've got this going on. Would it be of interest to Anthem?" One deeper interest of ours was the architecture and how they use AI and blockchain as a safe way to use consumer data to match them up to appropriate clinical trials.

So we decided to look more into it. And that led to a decision to move forward. So we're piloting that with Anthem in-house, with our employees. And then based on the results of that, we might conclude that it works, and we'll throw it out to a broader population. Or we might learn something from it. And based on all of that, they might become one of our ecosystem partners.

Kevin Pereau: Rajeev, where do you see the industry going in four or five years. Where does Anthem fit as a driver of disruption? If you're successful, what will the world look like?

Rajeev Ronanki: I think where we're headed is that Anthem will be a creator of an ecosystem that creates solutions for consumers. And in so doing, what we'll do in five, maybe push it out to a seven-year horizon, I would envision us being able to dynamically create an ecosystem based on real consumer data and deeper needs.

That level of dynamic creation using the resources is technically possible today, but there's no business model that requires it. If everything unfolds in the way that I think it will, that's where we're headed. It will perhaps be a world where you can essentially configure a set of assets on the fly, and people can either opt in or opt out. And that's probably how health care is going to work.

EXPERT OPINION: BRAD FLUEGEL, FORMER SENIOR VICE PRESIDENT, CHIEF HEALTH CARE COMMERCIAL MARKET DEVELOPMENT, WALGREEN COMPANY

In years past, most people viewed drugstores as nothing more than prescription-fillers. Patients took in handwritten prescription requests from their doctors, came back a day or two later to pick up their bottle of pills, and never visited again until it was time to refill the prescription or fill a new one. Perhaps that model was not entirely accurate, because customers also bought bandages, braces, deodorant, and maybe the odd bottle of witch hazel.

Today as you know, all that has changed. People are going to large pharmacy chains to not only get their prescriptions filled, but get flu shots, blood pressure checks, and advice from pharmacists. And those changes, while highly visible to anyone who has visited a pharmacy recently, are only the beginning. Retail pharmacies, which once occupied a position on the sidelines of health care innovation, are moving to play a wider and more important role in helping people lead healthier, more connected, lives.

Brad Fluegel recently retired from his position as Senior Vice President and Chief Health Care Commercial Market Development Officer for the Walgreen Company, where he was responsible for all commercial health care activities, including sales and contracting, biopharma relationships, retail clinics, clinical affairs, new service development and market planning. Walgreens currently operates more than 8,000 stores in all 50 states as well as the District of Columbia, Puerto Rico and the U.S. Virgin Islands. Sales in calendar year 2017 exceeded $87 billion.

Prior to joining Walgreens in 2012, Brad was an executive in residence at Health Evolution Partners. He also worked at WellPoint, where his responsibilities included long-term strategic planning, government affairs, corporate communications, corporate marketing, corporate development, international expansion, innovation and new business ventures. Earlier, Brad was Senior Vice President of National Accounts and Vice President, Enterprise Strategy at Aetna, Inc. where he was responsible for developing and executing strategies that expanded Aetna's position as a leader in the large employer market. Brad earned a master's degree in public policy from Harvard University's Kennedy School of Government and a Bachelor of Arts in business administration from the University of Washington. He is currently a lecturer at the University of Pennsylvania's Wharton School of Business and serves on the boards of many organizations, including the Metropolitan Jewish Health System (New York) and Performant Financial.

Kevin Pereau: How would you summarize your experience for our readers?

Brad Fluegel: I was at Walgreens for a little more than five years. I joined in October 2012. And before that, I spent most of my career working in health plans and health insurance. I headed strategy for Aetna, I ran strategy and external affairs for Anthem. I also was with the Harvard Community Health Plan, and then before joining Walgreens, I spent a year and a half or so in private equity.

Kevin Pereau: What big trends have you seen emerge over the course of your career?

Brad Fluegel: I think one of the things that is definitely a trend, but which has taken longer to be adopted than I would have expected, is the movement to focusing first on the patient as part of the health care journey. Historically, the focus has been built around physicians and their needs. It now needs to be built around patients.

That helps explain one of the reasons I joined Walgreens. I believe that the ability of consumers to have more convenient access points to care is a trend that people have been watching—one that everyone realized was going to happen everywhere. It is taking a bit longer to materialize than I anticipated, but I think that having patients served in convenient locations is a trend that is accelerating. That is why a number of pharmacies, including Walgreens, are becoming places that allow customers to do much more than just pick up pills.

Regarding other trends, let me mention telemedicine. People's ability to access care using their phones, laptops and tablets has been around for quite some time. I would have thought that it would have taken off earlier than it has. But it is now definitely accelerating.

Kevin Pereau: Do you think that thanks to telemedicine and other tools, individuals now feel they are more in control of their care—more in the driver's seat?

Brad Fluegel: Yes, I think that's true for most people. Customers' ability and interest in reading certain symptoms, in having access to more transparent information online, and to research their conditions is accelerating. And increasingly now, customers will be able to use their phones to schedule physical appointments and perform other functions. At Walgreens, we have a Pharmacy Chat function; if people have questions about any medications they are taking, they can chat with our pharmacy staff.

Kevin Pereau: How is a company like Walgreens using the growing amount of individual data that is available?

Brad Fluegel: There are a few issues to mention. We need to better aggregate the data that we have, the pharmacy data that we have, with other data related to what's happening with that patient.

When a customer comes into our store, for example, we know what medicines they are taking, whether those medications were purchased from Walgreens or from another pharmacy. But there are other things about the patient that we don't know. We might know that he or she has a chronic condition, because we know what drugs are taken. But if the customer is, say, diabetic, we don't know their glucose scores or A1C test results. And we don't know if they are having other medical issues that the pharmacy might be able to help with.

That is consistent with one of the questions that is being asked everywhere in the larger health care system, which is, how can we get and use a view of everything that is happening with that patient, so that any care provider, including the pharmacist, will be able to help that patient take better care of him or herself?

Kevin Pereau: Are we on the cusp of connecting patient data to pharmaceutical companies who make the drugs?

Brad Fluegel: Currently, I don't think that everybody in the system - whether it's an insurance company or a doctor or a pharmaceutical manufacturer - has enough information to know what's happening with a patient who is using a particular drug.

There are a fair number of programs, typically around specialty medications, that gather and relate information back to the manufacturer about what is happening for patients who are taking their drugs. I think that represents an opportunity for everyone in the chain of care to have more information about how those drugs are impacting on individual patients. But overall, I don't think that gathering that kind of information is something that's of great enough interest to health plans and manufacturers.

Kevin Pereau: Do you think we are moving toward a time when people and their insurers might not have to pay for drugs that have not helped them? That's kind of a thunderbolt.

Brad Fluegel: At first it will start with payers and insurance companies, but ultimately it may get down to individual patients as well. If I am dealing with a thousand-dollar deductible for a drug that doesn't work for me or my insured members, why should I be paying for it?

Kevin Pereau: Boy, that's a wonderful thing to say. To turn the discussion back to what Walgreens is doing, can you tell me how you are helping customers in the increasingly competitive environment of retail pharmacies?

Brad Fluegel: Both we and other retail pharmacies are going to become more full-service health care destinations where people can have a variety of things taken care of. We now have retail clinics in approximately 400 of our stores, and we are setting up urgent care clinics in about a dozen locations with physicians and nurse practitioners on staff. Lab services are being tested in our stores, so people can have blood drawn or urine samples taken.

Kevin Pereau: Patients' ability to go to Walgreens and have diagnostic tests taken is a very big thing.

Brad Fluegel: Our advantage is that we are already located close to where people live, work, and drive. We are better integrating delivery of medications to the end consumer, and to the health plan. The result will be an integrated experience, so patients will receive better care.

CHAPTER TWO

TELL ME WHERE IT HURTS

What is slowing our transformation to Digital Health 3.0, that we explored in the last chapter?

Frankly, physicians, hospitals, insurers and all the other big entities that dot the landscape of health care all contribute to the problem. Yet at the same time, they are also instrumental in trying to solve the problem. They aren't blocking progress because they want medical consumers to live with diseases and chronic conditions and die young.

They are part of the problem because they *have* problems. Or perhaps more accurately, because they are all suffering from one very large, overarching problem that we can call:

THE INCUMBENT'S DILEMMA

It is very difficult for organizations to disrupt their entrenched ways of doing things. They realize that they are already comfortable with their existing model. They profit from certain products, processes and services, so they don't want to change them. But what if those products are the very thing that must be changed? Organizations can never move ahead if they are unwilling to disrupt themselves at the most basic, seismic levels.

Let's take a closer look at why.

THE PHYSICIAN'S DILEMMA

Physicians attend the best medical schools they can, and take part in the most prestigious residencies. They learn the latest tips, treatments and tricks for helping us survive the most serious ailments.

Then they go on to see and care for patients. But let's face it, many of them would like to see those patients just once a year for a checkup. Certain caring, committed physicians like to see patients more often – often, patients who have chronic conditions that require regular monitoring.

But I think we can all agree that physicians generally see us when we're broken – when we're having problems, when something should ' be fixed.

In fact, physicians' classic business practice model is set up to see us when we have issues, and to not interact with us at all when we're healthy. They get paid when they see us. If you look at that situation from a holistic perspective, you realize right away that the caregivers who are most knowledgeable about how to keep us healthy only see us when we are ill, and rarely when we are healthy.

Much of the information and the data they collect about us centers around the one most troubling condition or disease that each individual may be suffering from and monitoring most closely. I tend to have high blood pressure, for example, so my physician checks my blood pressure and compares it to what it was when I had my last appointment. If I had diabetes, chances are my doctor would check my blood sugar levels, send me to have an A1C test that gives an idea of whether my blood sugar levels have been well controlled in the months since my last visit. And possibly discover, thanks to that test, that my condition has deteriorated.

The result is that we reflexively manage our conditions, or recover from flareups, or learn that it is time to adjust the medications we are taking, which have clearly not been up to the job of keeping us well since our last visits.

Because our physicians usually only consider what has happened in the past, they are often only able to help us to just live with our illnesses and conditions. If they were continually collecting and analyzing data about us, they could be helping us thrive with those conditions, not just survive from one day to the next. Most medical consumers, I believe, would pay more for that kind of attention and care. Some physicians, though certainly not all, might need to be reminded of that.

We need to be moving to a place where information about what is going on with our health is gathered continuously, not occasionally.

Another part of the dilemma is that most care providers are currently more than a little overwhelmed and confused by the information they are already receiving, even though they are not receiving enough.

There is a pervasive mindset in which care providers are wondering . . .

WHAT EXACTLY DO I DO WITH ALL THIS DATA?

In other words, physicians are often already overwhelmed with the quantity of data they are dealing with. And at the same time, all that data is not enough. We meet with our physicians and say to them, "Please look at this information and coach me to do the right things and influence me in the right direction." If we have good physicians who manage a reasonable patient load, they can do that. But in many cases, they cannot, despite good will and good intentions.

There is another way to frame the challenge that physicians are facing. They only make money when we are sick, not when we are well. There is a pressing need for health care to provide a framework that allows physicians to monetize the relationship with patients when they are not sick.

There is also a disconnect happening between my data and what my physician sees. In my case, for example, I am using digital tools to collect more and more data about myself – my exercise, weight, diet, blood pressure. I could feed that data to my physician and say, "coach me in the right direction."

That is where we're going, certainly. Interestingly, the Affordable Care Act drove a highly disruptive change that moved us toward a more value-based care model. But are we there yet? We have a long way to go, but a lot of progress is being made.

EXPERT OPINION: DR. THOMAS MUNGER, HEART RHYTHM DIVISION CHAIRMAN, MAYO CLINIC, ROCHESTER, MINNESOTA

How do physicians feel about the growing role that computers and data play in the quality of care they are able to provide? What has changed in the way data is used? How do they think the use of data is transforming their patients' lives, health, and wellness outcomes? And how do they view privacy issues?

It will come as no surprise that many physicians who care deeply about their patients care about those questions too.

Dr. Thomas Munger completed his undergraduate studies at the University of Vermont in 1980, where he majored in mathematics and minored in physics. The next year he entered that university's School of Medicine and earned his MD in 1984. He then did his residency at the Mayo Clinic, where his interests and passions led him to pursue work as a researcher, a caregiver to patients, and a teacher.

Dr. Munger has remained at the Mayo Clinic ever since, and now serves as Division Chair of the Heart Rhythm Division. He was awarded the Mayo Clinic's Cardiovascular Laureate Award in 2007. He was named the Cardiovascular Division Teacher of the Year in 2002 and has won numerous commendations that reflect his contributions to education, research and patient care.

DR. MUNGER ON COMPUTERS AND MEDICINE: A 1983 PERSPECTIVE

Dr. Munger began to consider the use of computers in medicine while he was still in medical school in Vermont.

"On one occasion, I think in my third year of medical school, we had Dr. Larry Weed come visit us from Case Western Reserve University," Dr. Munger recalls. "He was a big proponent of putting a computer on every desk, right in front of every physician, to be used to try to figure out diagnoses."

In that class, Dr. Weed went to a blackboard and drew a big circle and said, "This is the medical knowledge that you possess, or what the valedictorian at the Harvard Medical School knows."

Dr. Weed then drew a bigger circle around that circle and said, "And this is what you need to know." And then he said, "This is why we need the computer in health care."

Dr. Munger was no stranger to computers. He had used them a lot during his undergraduate years. "We were writing Fortran in about 1983," he recalls. "From a medical point of view, we used a little computer, it was a touch screen computer system, in the main medical building up in Portland to do order entry. Nothing else was on it. It had a cumbersome touch screen but that was the first time I used any computer in health care. But when I understood what Dr. Weed was telling us, I saw the enormous potential that computers could have in health care. But I also saw a roadblock. How would we ever get all that knowledge, and all that information, into a computer system that everyone could use? I'm not sure we've ever gotten there, after 30 years."

FROM PNEUMATIC TUBES TO COMPUTERIZED PATIENT RECORDS

When Dr. Munger arrived at the Mayo Clinic, he discovered that a rather sophisticated, while antiquated, system of delivering patient records was in use.

"The idea of an integrated medical record had been conceived before the computer existed," he explains. "Years earlier, Dr. Henry Plummer had worked with the Mayo brothers and they decided to create a way to take paper records and integrate their use on the medical center campus, so that the surgeons and the internists and the radiologists and everyone could always see a patient's records. They developed a system of pneumatic tubes that were used to deliver a patient's records to different departments. So, there was an independent record that traveled with each patient."

He recalls that the system worked remarkably well. But as you would imagine, there was a need for a more sophisticated electronic system - one that would not only deliver paper records, but capture patient data in a central place. So by 1995, he and his colleagues were moving patient records over to the Mayo's own Electronic Medical Record (EMR) system. "The challenge was to integrate a hundred years of paper records into the digital records," he says.

"And it basically never happened. Even now, if I need records from the 1940s or 1930s, I still have to go back to the paper records. And there are other challenges. We have two other large groups - one in Arizona and one in Florida - and the goal is to have shared access to electronic records everywhere. We made a decision at the enterprise level to switch from the EMR system that we developed in Rochester to Epic, a vendor. So, we're working through that. I think the idea at the enterprise level is that there must be something that everyone can see around the country. While physicians still won't be able to see medical records from non-Epic users, I think you'll see that when a patient comes from Brigham and Women's, for example, since they're on Epic, we'll be able to see their records. So that is a good step. Again, the challenge is trying to integrate records amongst lots of different centers in the country. We're moving the whole process over to Epic. I think we just happen to be in a three-year project."

Previously, it was up to the patient to collect a variety of medical records. If he or she had multiple health issues and was seeing several different physicians, it was necessary to contact each of those providers, request records, organize them, and make them available to everyone. Some very fastidious patients collected all their records and brought them all to each physician they saw. Some of them even organized their records on their own computers. But it was very inefficient and haphazard.

"And then," Dr. Munger adds, "there is the problem of getting all those paper records into our EMR by scanning and organizing. And many times, the paper print-out of those records is very cumbersome and poorly organized. It doesn't meet the needs of the patients, insurers, pharmacies, or other entities."

HOW DR. MUNGER SEES SECURITY CONCERNS

The security of patient data is a major concern, both for patients and care providers.

"I think it's a huge thing," Dr. Munger says, "and it's not actually standardized across the country in any fashion, as you would expect it to be. I think there are 4,000 possible hospital systems at this point in time, and God knows how many more additional outpatient clinics."

And now another concern is emerging: the need to protect data and systems from hackers. In one recent case, they were able to shut down a hospital's emergency room and demanded a ransom, paid in Bitcoin, before they would turn it back on again. And there are still more concerns.

"Another major concern," Dr. Munger explains, "is the possibility that individual patient devices and data can be hacked. Being able to wirelessly communicate with devices like pacemakers and implantable defibrillators is important. We can pull data off those devices on a day-by-day basis to see how patients are doing. But it is theoretically possible to hack into that data and cause havoc. Security issues are always growing and evolving, and we need to be equally creative about implementing systems to limit access to patient information."

Every kind of implantable device that is about to be used with a patient is analyzed by the Mayo Clinic and subjected to an internal security analysis and review. Another security concern centers on the number of emails that physicians and other personnel receive every day. To help them remain vigilant about possible phishing emails, the clinic sends out reminders about phishing and fake emails that resemble phishing efforts; if staff members make the mistake of responding to them, they are reminded of appropriate security protocols that they should have observed.

IMPROVING PATIENT OUTCOMES

When Dr. Munger first began to practice, patients typically visited their physicians every three months. In those times, data was updated in a non-continuous way. Today, electronic data is gathered continuously.

"Once wireless technology came out with implantable devices and patients had modems installed in their homes and similar technology was put in place, things changed dramatically," Dr. Munger explains. "We started getting data on a daily basis."

One big issue now is monitoring and triaging that incoming data, so emergencies can be recognized if they occur, and preventative care can be initiated. The Mayo Clinic, has a staff of technicians who constantly watch incoming data and act quickly if, for example, it indicates that a patient is having a cardiac event.

"They're constantly monitoring about 100 patients," Dr. Munger says, "and they have a protocol with which to escalate it to the next level, to the coronary care unit or attending physician."

Many companies, Dr. Munger states, are developing software that can monitor that incoming data and issue alerts if a patient should be advised to go to the emergency room, contact his or her physician, or take other appropriate steps.

WHEN DATA-GATHERING IMPEDES DATA ACCESS

Now that patients are gathering data on their own, other concerns arise. Dr. Munger explains, "Patients are gathering more of their own data today, which is positive. But in the process, is important and possibly critical data being withheld from care providers who should be monitoring it?"

When patients arrive at the Mayo Clinic for treatment, they are asked about all the points where their data is being collected and reported, so the Clinic can both gather it and take appropriate steps to counter any security risks.

FUTURE CHALLENGES AND TRENDS HE SEES

One big concern is the need for more widespread data-sharing across hospitals and physicians. If a patient who is receiving treatment at one hospital visits the emergency room at a hospital located far away, for example, impediments still exist in accessing that patient's data immediately and efficiently.

Another obstacle is that even when large databases of patient data have been amassed, it is an immense challenge to search through it, find what is pertinent, and discover patterns that indicate that treatment might be required. What medications is a patient taking, for example, and are they likely to cause unhealthy reactions? Making such determinations is still largely up to individual physicians who review records during patient visits.

Yet as more data is amassed and better organized, major new benefits lie ahead for patients.

Dr. Munger sees one major trend on the horizon: the integration of patients' genomic data into their medical records. If a patient has a genetic and familiar predisposition for a particular disease or condition, that information can be integrated into his or her records, where it will trigger alerts when certain types of tests become available, when certain drugs should be avoided, and more.

"We know that genotyping, which has been out there for 20 years, can have a major impact on patient outcomes," Dr. Munger states.

"When it is integrated into patient records, it will add up to a whole personalized approach to medicine."

THE DILEMMA FOR INSURANCE COMPANIES

We think of insurance companies as big, profitable money-making machines.

Like physicians, insurance companies are confronted with the need to change. But because they are infected with the Incumbent's Dilemma too, their ability to address certain critical problems has been slow to leave the station, and not without missteps.

In most other industries, companies build their products from the ground up, to meet consumers' needs. Insurance is one industry where that hasn't exactly happened. The range of products confuses customers more than in any other sector.

Complex big insurers have developed a remarkably complicated range of products. No question, they design a bunch of those products right. But they are constantly evolving, never static. Those companies develop and revise their policies and products continually, then withdraw them if they fail to generate enough profit.

As an end user, you have seen that happen if you've ever gotten one of those confusing notifications that says, "Your Policy A no longer exists, but don't worry because we have automatically enrolled you in our wonderful new Policy B." Just when you think you have your insurance coverage running steady, everything changes. All kinds of actuarial analyses are taking place behind closed doors at your insurance company, but no one takes the time to explain to you how decisions affecting you are made.

In a way, insurance companies face the same problem that carmakers face when conceptualizing and selling their products.

As an analogy, let's look at BMW. Let's say that you're a car buyer for whom cost is no object, so you buy the highest end BMW you can find in the dealership. Based on that car, you conclude that BMW makes the best car in the world. But if you are budget-conscious and buy a low-end, entry-level BMW, you drive out of the dealer's lot and soon you are thinking that your new car is just another vehicle.

Insurance offerings are a lot like those cars. If your employer purchased costly medical coverage for you and you're getting high-end health screenings and all kinds of other "bells and whistles" extras, you feel that your insurance company is the very best there is in the whole wide world. But if your company purchased a low-end plan from that same provider, you probably don't have a very high opinion of that insurance company. You're always being told that certain diagnostic tests are only 50% covered by your plan, that certain new drugs are not yet reimbursed, that you owe money because you visited an out-of-network specialist. You know exactly what I am describing.

Part of the problem lies in the way employers purchase insurance, and in the choices that insurance companies have made in how they package and sell their products. Typically, they try to sell their products to companies because they make more money selling plans than they do by knocking on your door and selling you an individual policy. Big companies with thousands of employees – the companies with the deepest pockets - they reason, can be sold full-featured plans and policies.

As a result, the real-world needs and desires of individual consumers like you and me are sunk two levels down in the process when new policies and plans are designed. We do not always get what we need or want, because we are often not being considered by the companies selecting our plans.

The insurers' dilemma is often one of complexity. They are committed to serving their members. But will pursuing membership growth help them deliver better health care services? How are they helping medical consumers like me and you?

The consumerization of health care has been especially disruptive for insurance companies. They are adapting and, in some cases, lead the innovation efforts.

PHARMACEUTICAL COMPANIES

Big pharma is facing a conundrum that is built into the way that they develop drugs and sell them to consumers. Do pharmaceutical companies understand how they develop and sell their products to consumers? They do, but again, the individual consumer is not at the epicenter of their business model.

They collect data on the potential market size for their current and future medicines. They can tell you, for example, the aggregate size of a new market for a new drug for diabetes, hepatitis C, atrial fibrillation, and so on. They collect data on where the potential buyers for those medications live, and where they buy their medicines. They also have data about which retailers are currently selling the pharmaceuticals that they already make – and they can use all that data to project which retail pharmaceutical chains will sell the most of their new product, how much of it they will sell in which areas of the country, and so on.

Thanks to large clinical studies, they can even make projections about what percentage of a particular population might be positively affected by the introduction of a new drug. In other words, their business depends on developing and analyzing aggregate statistics.

What they can't tell you about are customers like Heidi, John or Paul. They are only beginning to collect individual data and interact with consumers on a one-to-one basis. At the end of the supply chain the consumer – you and I – fills a prescription, begins to take it, and reports back to a physician only if something does not work well. There is little back-and-forth communication, and no data collection or monitoring once the consumer starts to swallow those pills or take those shots. However, systems are emerging that can communicate digital data providing deeper insight into the drug's efficacy.

CVS, Walgreens, and other retail pharmaceutical stores now maintain long-term electronic records of the medications that customers use. In those same stores, the shelves are stocked with a range of devices that can collect individual data. There are smart scales that record and track individuals' weight and communicate to tracking platforms. There are connected meters that measure blood glucose levels and blood pressure. There are even Wi-Fi enabled electric toothbrushes that track the amount of time that their owners spend brushing their teeth daily.

Everyone in the pharmaceutical supply chain – from the researchers who develop drugs to the end-users, with every other entity in between – is using digital health assets to put the pieces together and connect the dots. Consumers benefit because this connects them with the companies that develop medicines without disintermediating anyone else in their value chain.

EXPERT OPINION: DR. ALAN PALKOWITZ, DRUG DISCOVERY SCIENTIST AND HEALTH CARE FUTURIST

How much does the man or woman in the street know about pharmaceutical companies? Certainly, those people know something. They know that drugs cost an awful lot, and suspect that pharmaceutical companies are profit engines that sit around waiting uneasily and charging high prices until their products become generic.

But many consumers, if they have had the wonderful experience of being cured by a drug, might feel somewhat differently about pharmaceutical companies. If they have a family member or friend whose health has been restored thanks to a wonderful new drug, they have a more positive attitude toward the companies that make drugs.

Few people, however, seem to realize that there is a deeply moral aspect to what pharmaceutical companies are doing. Some of their scientists have devoted their lives and remarkable intelligence to saving lives, restoring health, and defeating disease.

Dr. Alan Palkowitz recently retired from Eli Lilly and Company, where he served for 28 years. His most recent leadership role was Vice President of Discovery Chemistry and Research and Technologies. At Lilly, he supervised the work of more than 500 scientists and was responsible for small molecule (oral medicines) drug discovery for diseases that include cancer, diabetes, immunology, pain and neurodegenerative disorders.

Dr. Palkowitz's love of chemistry and biomedical research began early in his life and his career has kept him at the leading edge of pharmaceutical research. As he enters a new phase of his career, he stands among the most knowledgeable medical researchers in the world.

Kevin Pereau: Medicinal chemistry and drug discovery is a highly specialized field of research that requires remarkable ability, experience and skill. How did you decide to dedicate your life to it?

Alan Palkowitz: My entry into research and the pharmaceutical industry was shaped by my undergraduate education at U.C. Berkeley. I was very interested in chemistry and in my second year, I started studying organic chemistry. And I really enjoyed it - organic chemistry could be called the chemistry of life, because it is fundamental to all living beings. I was inspired by a professor I had at Berkeley, Henry Rappaport. In the year when I started in his lab, he was focusing on structural modification of opioids, like morphine. Part of his research was trying to modulate their biological activity, with the goal of making painkillers non-addictive. It is an issue that is still important today.

I was working with graduate students in his lab, many of whom were thinking about their career direction. Several were going into the pharmaceutical industry, because that was where many job opportunities existed for scientists who could design and make molecules, which is a central aspect of drug discovery. In other words, scientists who wanted to find ways to create molecules that function as medicines. The idea of using the fundamental rules of chemistry to help people really excited me.

So that pretty much set me on my course. I wanted to further develop my skill as a chemist to prepare for a career in pharmaceutical research and so after Berkeley, I went to MIT for my doctorate. And after that, in early 1989, I began to look for job opportunities in the pharmaceutical industry. I accepted a job at Eli Lilly and that began my career. I entered as a bench-level chemist, working on therapies for cardiovascular diseases.

Kevin Pereau: How did you make the transition from chemist to a leader in Lilly's enterprise and business activities?

Alan Palkowitz: Early in my career, I had a fairly rich experience as a laboratory scientist that was excellent grounding in the challenges and opportunities in drug discovery. I then gradually moved into managerial positions and began to take a larger interest in the broader dimension of the company, thinking about research in the context of business and the enterprise. During my time at the company, I saw a tremendous transformation from the early 1990s until today. Eventually, during my last 11 years at Lilly, I took on the position as Vice President of Discovery Chemistry and Research and Technologies, and I was responsible for small-molecule drug discovery.

I became very involved in building our pipeline of innovative medicines. And with that came a larger responsibility for the process of taking a therapeutic concept and converting it into a molecule that could be tested and developed for use in humans. And hopefully, one that could become a drug. That involved managing a large portfolio of projects, dealing with scientific complexity, differentiating our work from competitors, creating intellectual property, as well as making sure we were continually investing in promising science and technology that would keep us on the cutting edge of creating novel medicines.

Being with a company that is at the forefront of the race to find medicines has been tremendously fulfilling. I recently retired, at the end of last year. I am now planning to explore new types of opportunities in biomedical research.

Kevin Pereau: This is maybe too big a question, but what major trends have you seen during your years in the pharmaceutical industry?

Alan Palkowitz: Let me generalize first. Certainly, when I entered the industry, there had been a tremendous amount of progress in the previous decades in many disease areas. For example, in cardiovascular, statins and blood pressure medicines were dramatically changing the management of heart related disease. Additionally, many versatile options existed for the treatment of infectious diseases and was an area of major emphasis among many pharmaceutical companies. The control of diabetes, while still a growing problem today, had been advanced due to the availability of biosynthetic insulins. So that era was a time of tremendous growth and many companies began to venture into new areas of further unmet medical need. But during my time at Lilly, I think the trends have been marked by tremendous advancement in our understanding of disease and medicines at a more molecular level. And there have been more and more insights in developing new types of therapies and interventions—getting beyond the symptoms and focusing on the underlying causes of certain diseases that may someday lead to cures.

When I think about the past 30 years or so, we have gone from a reductionist approach to more of a knowledge-based approach, one that is now being informed by tremendous clinical learning and an increasingly large and accessible volume of data from patients. We are developing a deeper understanding of why some patients respond, and others don't respond, to certain types of therapies, and why some therapies can be safer for subsets of patients in a given population. So we are peeling away layers to get a more fundamental understanding of diseases and how to address them. That has accelerated innovation, driven by novel insights and informed experimentation. And also, we are arriving at more sophisticated methods to visualize the potential interactions of molecules with target proteins, which gives us insight into how to attack disease and design better medicines at the molecular level.

Biotechnology is also now playing a bigger role in expanding the breadth of medicines that are available for many diseases. Several years ago, we had only a handful of biomolecular therapies like insulin or growth hormone, that mimicked natural proteins in the body. Today, thanks to the advancements in biotechnology research and development, the industry has created therapeutic antibodies and other complex proteins that have transformed the treatment of disease. In oncology, for example, we now have immune checkpoint antibody drugs that have dramatically changed the treatment paradigm of many tumor types.

For patients with rare single point genetic abnormalities, like hemophilia, we are seeing innovative gene therapy approaches progressing in development. These trends are exciting and could lead to curative paradigms for some rare diseases.

But even so, there is still a long way to go - a lot of opportunity ahead of us. We have made great progress in oncology and the ability to prolong life expectancy for some cancers has been great, but there are many tumor types where solutions are still years away. We still don't have effective treatments for Alzheimer's disease and few viable pain medication options. So the future will be informed by a greater basic understanding of disease and dependent on the collective work of academic and industry researchers around the globe to create new avenues of productive discovery.

Kevin Pereau: And looking ahead?

Alan Palkowitz: When I look a little bit forward, I think that the intersection of drug discovery and digital information will become much more important to improve overall patient outcomes. You can go all the way from discovery to a phase-three trial and find out the drug you are testing didn't work, or wasn't safe, or wasn't better than standard of care. This is extremely costly both in terms of dollars and opportunity. So that creates a challenge, but also a great opportunity. One of the burdens we have in our industry is that we must continuously force the innovation frontier, the innovation curve. We need to focus on the best possible disease hypotheses, accelerate potential medicines in research and development that will have a large impact, minimize clinical failures, and find ways to improve drug access and compliance for patients. Continually harvesting knowledge and critical information from many sources of data along the value chain will be a key variable to addressing these challenges.

We must consider how we use patient data, clinical data, and research data to better understand the drugs that we are developing, whether they will have the impact we envision, and whether they will be safe. We need to think about how the drugs we develop will reach the patients who will derive the most benefit from them. For example, there are genetic fingerprints that suggest how certain patients are going to respond to certain drugs and how to administer those drugs more precisely. The ability to do that is central to precision medicine initiatives that maximize the impact of medicines and avoid treating patients that we know will not benefit. This is already happening in diseases such as cancer.

So, precision health is a very important trend and value proposition. One of the things that is going to drive it will be the ability to harvest learning and data that is associated with individual patients, along with the data from the many clinical trials that are ongoing today. That will yield insights into how some patients respond and some do not, and to help insure that the investments that drug companies are making will actually result in better outcomes.

In effect, that means trying to insure that better ideas will result in better drugs that will be valued by payers and providers, and which will ultimately have a profound impact on improving patient's lives.

Kevin Pereau: You mentioned data. And since that is the subject of this book, can we talk about that again?

Alan Palkowitz: So that is where convergence is taking place, the place where the progress that has been made can now be amplified. Suppose we are able to use data to simulate clinical trials, or a trial for one individual, and understand whether he or she will respond or not respond to a particular treatment.

With data and modeling, we should be better able to understand what the transformative effects of a drug will be over a span of years. We should move toward developing predictive models and use them in the early research and development space before we ever study drugs in humans, so we better understand the potential outcomes.

While likely to not be definitive, this approach may better inform the design of actual clinical trials as well as reduce costs of development. Again, I think the collective body of data being generated from so many places - from treatment, clinical trials, understanding the genetic basis of disease - is going to converge in some pretty special ways. Today, we are only beginning to scratch the surface.

Kevin Pereau: Does government regulation impede or help the adoption of new technologies?

Alan Palkowitz: I think everyone involved in the ecosystem has a vested interest in advancing health care. I think we all do, and I don't think government has been an impediment. The FDA has been more proactive in its initiatives to speed up the approval of drugs that will have a positive impact on patients. The FDA also has a very important role, which is to ensure that we're addressing patient safety. So, I think that is a very good relationship that will continue to evolve.

I think there is greater and greater sophistication in the FDA, for example, in understanding how to partner with the innovators in the medical space. It has been very exciting in the last year, really a breakout year in the development of new drugs. Forty-six new drugs have been approved. So that is a sign that there is a tremendous amount of progress being made, both in the industry and with regulators, to bring high-potential therapies to market as quickly as possible.

To get back to digital health, the pharmaceutical industry is increasingly interested in utilizing data to optimize the care of patients. And I think that making patients better informed and more in touch with their health care will ultimately improve their compliance, which is necessary to maximize the effectiveness of medications. Coming from pharma, I believe that medicines are only going to be effective if they are properly and consistently used . . . if they are matched to the individual patient for maximum benefit. And some of the digital health tools that are coming will specifically address compliance. We know, for example, that certain therapies can reduce the incidence of cardiovascular disease, but only if they are properly utilized every day.

That would be a very valuable outcome when you think of the overall health care problem.

That is why at Lilly and many other companies, there is emphasis on approaches to connected care in applicable settings for disease management. For example, there are devices that continuously monitor blood glucose, and can administer insulin on demand, based on what is happening with the patient. And then you have information that is being collected and delivered to health care providers, so that medication can be adjusted when need be. Integrated approaches like those will be transformative.

I also think that there are now some big opportunities in chronic diseases, where there aren't obvious signals when treatment is not effective. If you have diabetes and your blood sugar becomes too high or too low, you will know because of obvious symptoms. But if you have high blood pressure, unless you check your blood pressure, you might not even know that it is elevated. But if temporal data can be collected and shared by a device that assures that patients are aware of the need to take medicine, that can be very powerful.

I do think though, that there is oftentimes a tendency to over-hype, to begin to extrapolate too far about what might be possible with information and associated technology. Of course, many things can be done with information. We will soon have more phone applications that will utilize artificial intelligence to gather and contextualize information as well as suggest action. You already have people wearing Apple watches that are reporting back to their physicians, but I think the key thing is that it must somehow lead to a meaningful action and outcome, and not just an information exchange or collection. And so, I think that as digital health evolves and many more ideas surface, some will emerge as important and impactful, while others will not be scalable in a meaningful way.

Overall, we are looking at ways to improve results and reduce the burden on health care systems. It is an exciting time ahead to see how data can be used in the most productive ways to make this possible.

Kevin Pereau: Let's talk, if you don't mind, about a story that has been making the news only recently - the merger between Aetna and CVS. Do you think that their merger serves as a predictor of other events to come?

Alan Palkowitz: In a sense, I think that the merger gives us a small glimpse of how different entities on the health care landscape will begin to cooperate in some very interesting ways.

We begin to see ways that different systems can work together. CVS is normally a place where people go to pick up their prescriptions. CVS stores also now have clinics where customers, in cooperation with their insurance providers, can enter into more of a partnership with CVS.

We are already noting that instead of going to emergency rooms when they have colds, for example, customers will visit CVS in-store clinics. That offers a way for individuals to go where they will find the most effective treatment option, in real time and at the lowest cost. That will be tremendously powerful. It will allow insurers to provide the most effective coverage for their customers, as well as for different providers within the health care community. It will eliminate waste. If you are a physician for example, you'll be better able to focus your high-end skills on the patients who need them the most, not spending your time treating patients whose minor ailments can be dealt with by a nurse practitioner or at a walk-in CVS clinic, for example.

I think all the things we have been discussing have the potential to work together very proactively and have an impact. All the information that is being gathered will begin to serve customers more effectively with their medications, with their health care providers, with their insurers, and more.

Ultimately, progress will depend on new ways we find to contextualize information, integrate technology with devices, and maximize the impact of diverse contributions made by pharma, insurance companies, care providers, and others to improve health. It's a broad, broad spectrum, and we are entering a very exciting time.

THE HOSPITALS' DILEMMA

Providers are also being presented with transformational challenges. They will need to change in the interest of serving customers in the coming digital age of health care.

For hospitals, the central piece of the holdup is the fact that hospitals make money from filling beds. That, even in our fast-changing world of health care, remains one of their leading metrics.

Hospital directors are ultimately judged on their ability to keep the beds filled. They lose their jobs if they can't do that. And when hospitals conduct searches to replace the directors who couldn't get the job done, they look for new administrators who have built track records of filling beds at other institutions.

When hospitals are trying to fill more beds, they typically focus on ways to improve the patient experience for the patients who occupy them at any given time. There is nothing intrinsically wrong with that focus – it is a way to seize that proverbial "low hanging fruit" and make positive changes more quickly. But while they are seizing those grapes, bananas and apples, they tend to overlook larger, more pressing priorities that will produce more long-lasting change that will be of greater benefit to consumers.

What will shake things up and provide fundamental change? To put it succinctly:

HOSPITALS MUST FIND A WAY TO BE RELEVANT WITH CUSTOMERS WHEN THEY ARE NOT IN HOSPITALS.

Granted, that is happening more and more. Hospitals are conducting screenings for a variety of conditions, hosting special events that support people who are suffering from different ailments, and conducting educational seminars for children and other members of their communities.

For many of us, hospitals are not the happiest of places. We have unpleasant memories of trips to the emergency room, of recuperating from operations, of stays that went on too long . . . and worse.

But that is changing. Hospitals are no longer places you go to only for emergencies or surgical procedures. In new and remarkable ways, they are taking a role in keeping us healthy. They are providing ongoing services in areas like nutrition, exercise, and the monitoring of diabetes, heart disease, and more.

Data plays a critical role in enabling these changes. As the role and usefulness of data expands, what will hospitals do? What will they look like in the future?

EXPERT OPINION: TOM MARTIN, SENIOR VICE PRESIDENT AND CHIEF STRATEGY AND INFORMATION OFFICER, EVERGREEN HEALTH

Tom Martin is responsible for information technology, strategic planning and business and relationship development for Evergreen Health in Kirkland, Washington. Evergreen Health, a two-hospital system with an annual revenue approaching $700 million, is one of the largest and most innovative care providers in the Pacific Northwest. Evergreen's hospitals employ 300 health care professionals and are affiliated with almost 400 physicians in their integrated care network. Evergreen also operates one of the largest hospices in the region.

Tom came to Evergreen in 2006 from University of Washington Medicine, where he served as Chief Information Officer for the UW Medicine organizations, including the UW Medical Center, Harborview Medical Center, School of Medicine and all Practice Plans.

"They were writing software in the early nineties, and I went there as director of system development," Tom explains. Previously, he worked for Andersen Consulting in their state and local government practice. He earned a Bachelor of Arts degrees in mathematics and business administration from the University of Portland. And in 2014, he was recognized with a *Puget Sound Business Journal's* 2014 Chief Information Officer Award, in the category of "businesses with more than $1 billion in annual revenue."

TOWARDS A NEW STRATEGY FOR BUSINESS DEVELOPMENT IN THE HEALTH CARE FIELD

How do Tom's varied skills in technology and business development come into play in the evolving world of digital health? And what can he tell us about larger trends in health care?

"Health information is complicated," he says, "and it has been challenging to get the same levels of scale and benefit that you've seen in other industries. It has also been a challenge to arrive at the level of investment and benefits that have existed in other industries."

For such reasons, Tom says, investment and innovation in the health care industries have lagged somewhat behind those in other industries. He also has observed that government regulation has discouraged entrepreneurs from becoming as active as they have been in other industries. "Regulation has caused them to not be interested because of the steep learning curve and the fact that profitability and returns are just not there in the same way."

"We've always had the vision to do pretty amazing things with technology," Tom states. "We spent a long time, from the early 1990s until today, getting people to buy into the concept of building an electronic health record. Now people can begin to see value in that, and to derive value from it. We're not where we need to be from an interactive perspective. But with the amount of data that is now available, we're going to be able to engage consumers more. It is going to be exciting."

THE CHANGING FACE OF THE PATIENT EXPERIENCE

In the past, hospitals typically interacted with patients when they were inpatients, or outpatients who were using hospital facilities. Now the scope of those interactions is expanding. Patients tend to notice that change because they are turning to hospitals for a wider range of care options than they once did.

"The technology enables that," Tom says. "One reason is that there is more accountability from a financial perspective. Hospitals are graded, if you will, based on patient experience and outcomes, and so financial incentives are starting to drive the move toward more services too."

FROM "HOW WAS YOUR EXPERIENCE?" TO "HOW ARE YOU FEELING?"

Time was that hospitals asked patients about their level of satisfaction with specific hospital stays and hospital-provided services. Hospitals still communicate with patients about those experiences. But Tom points out the range and variety of communication between hospitals and patients is increasing. More information is flowing in.

"Digital tools allow us to have a more comprehensive interaction between visits, if you will.," Tom explains. "And we are able to get a better sense that people are feeling better. The level of interaction increased first with patients who had undergone orthopedic procedures. We were able to monitor how they were feeling, how quickly they were returning to normal activity, and how consistently they were going to physical therapy. And now, thanks to digital monitoring, we can intervene if they aren't."

And this increased level of monitoring and connectivity is expanding into nearly all areas of patient treatment and care.

"With orthopedics," Tom explains, "it has been about providing prompts to get people to do their physical therapy, to be more involved and compliant patients, and to take an active role in reporting their progress. It's more proactive than just monitoring."

Now Evergreen Health is launching pilot programs to increase data-gathering and monitoring for patients who are losing weight and who are dealing with conditions that include COPD, chronic heart failure, diabetes, and high blood pressure. In addition, a program is in development for women who have developed high blood pressure during pregnancy. And nutritionists and other health care professionals are becoming affiliated with Evergreen and offering care.

"There will be other forms of home monitoring that we do," Tom says. The result, he predicts, will be that patients will become more proactively involved in taking ownership of their own health care.

INCENTIVES FOR WIDER SHARING OF PATIENT DATA

Are we entering a new world where patient data will be more widely shared between physicians, hospitals, insurance companies and other entities in the health care landscape? Tom believes that a freer exchange of information will evolve. Yet from his perspective as a businessperson, he understands that there must be business reasons for those changes to take place.

"The government's requirement under meaningful use to really push data-sharing has helped and driven the evolving standards of health," he says. "And I think big consolidations within the industry and big partnerships have served as business reasons for sharing to happen and have driven real business value."

Trust and shared experience also play a role. Tom points out that insurers would like to have wider access to patient data, and that the level of trust between care providers and insurers has grown. "Yet as you might expect, we have to put protections in place where patient data is requested."

And patient attitudes are changing about the privacy of their data.

"I think this is interesting," Tom says. "People now expect care providers to know a lot about them. Yes, there are pockets of people that are really concerned about privacy, while others are starting to say that their data is just part of the digital world. Yes, some people worry that their data will get into the hands of the wrong people. But as you might recall, people had similar worries about their financial information about a decade ago."

So attitudes are changing.

BIGGER TRENDS THAT TOM MARTIN SEES

Where is the use of digital patient data going? How will it help patients experience better care? Here are some trends Tom points to:

- The predictive use of data. Care providers will be able to use data to engage in more predictive activities. "We will be able to monitor sets of populations for different problems and make smarter decisions about how we can help," he says. "We will be in a position to say, `Hey, something is up here.'"
- Intervention. Care givers will not only know when patients are experiencing problems or have needs, they will be able to take proactive steps to contact those patients and deliver care, even when patients live far away. "The care can be less invasive than going to a clinic or a hospital," Tom predicts. He also believes that thanks to emerging technologies, a wider range of actual care will be delivered remotely. "Sometimes, the intervention will be such that we will adjust what patients are getting as part of their care, and those patients won't even know we made them."
- An increased use of data in research. When data is collected more widely, it will find wider application in research. "We will be able to tie in lots of different environmental data and

make correlations we couldn't before," Tom predicts. Is pollution lowering the life expectancy of people in one part of the world or the country, for example? What affect is pollen having on public health? "We are going to be able to have more data to analyze questions like those and understand."

THE EMPLOYER'S DILEMMA

Most organizations today are aware of the link between good health and profitability. They understand that healthy employees are the most productive workers. They log fewer sick days, perform their jobs better, and are more likely to remain employed than are workers who are ill or suffer from debilitating conditions. The overriding concern among employers is what they perceive as the high cost of providing health care. At present, few employers understand that connected health care has the potential to dramatically reduce the cost of doing business, not increase it.

That helps explain why employers are committed to making cost-effective decisions regarding the health care benefit plans that they pay for and provide to their workers. The biggest challenge they face is to take a more active role in creating programs that keep their employees healthy, not only to address illnesses and conditions after they have begun.

A change is already starting in this area, as more companies fund not only medical plans, but offer programs to help employees maintain healthy lifestyles, quit smoking, lose weight, and prevent a growing range of health problems before they occur.

It takes a village. Employers play a big role when they realize it is their job to keep their employees healthy. It all boils down to the same issue that physicians, insurance companies and the other players are also facing—the need to keep medical consumers well, not fix them after they become broken.

Employers can serve as conduits that provide data to insurance companies about employees' health, patterns of exercise, and other factors. As we evolve toward the age of Health Care 3.0, their role will expand. Look for corporate wellness programs that reward health lifestyle choices.

EXPERT OPINION: RICHARD LUNGEN, FOUNDER AND MANAGING MEMBER, LEVERAGE HEALTH SOLUTIONS

In years past, the world of health care often seemed to be populated by a number of entities that were having a difficult time working together. Like an orchestra that was made up of musicians who were each playing parts of different symphonies, there seemed to be little order - and certainly not much harmony.

What would happen if a good conductor stepped up, handed out the right musical parts, and began to conduct and coordinate everything? What would happen if all those different players began to play in harmony?

And what would happen in the world of health care if that same kind of conducting organization stepped up and coordinated what all those different players were doing? What kind of experience would all those players have? And what about the people in the audience?

Something similar is worth working for in health care.

Leverage Health Solutions describes itself as, "The market leader in health care growth services focused on health care strategic business development. We concentrate on delivering best-in-class solutions by way of our Portfolio Companies to the health care marketplace, with an emphasis on payers and health plans, based on unique and unparalleled industry experience. Our team excels at understanding the needs of health care payers as well as emerging trends in the vendor community. All health care constituents are experiencing significant change and we are ideally positioned to offer proven results to dynamically impact all aspects of the industry. Our efforts result in measurable results encompassing profitable revenue, operational savings and market facing enhancements for payers delivered from our best-of-breed vendor solutions."

If that statement makes Leverage Health sound like the Swiss Army Knife of health enterprise, that is not inaccurate. Richard and his team do not specialize in just one aspect of health, but aim to improve processes and experiences for care providers, insurers, employers, health systems, pharmacy benefit managers, as well as end-users.

Here are some excerpts from a recent conversation between Richard and myself.

Kevin Pereau: What led you to start your unusual, innovative company?

Richard Lungen: I was working for a health and life insurance agency in New York state, where I learned the importance of dealing with local employers and their employees about their health care plans. In 1992, I was on the ground floor of a small, local third-party administrator who administered self-insured benefits for local employers. We went on to create a national provider network called National Preferred Provider Network. That organization was my entry into the provider network industry, which is still a big part of our passion at Leverage Health. Several of our companies are attached to the network side of health care, which we believe is the most important asset owned by health plans.

I founded Leverage Health in 2007 to enable highly qualified, innovative vendors to accelerate their growth into the health care marketplace. My co-founder and I saw an opportunity to become an integrated partner to highly promising companies, which we call our Portfolio Companies, and to help them grow their revenue and client base faster than they could have done themselves. And we have been doing that for over 11 years.

Kevin Pereau: What kind of companies do you work with?

Richard Lungen: All our portfolio companies provide some form of innovative and unique solution, service or technology - predominantly to health plans, but also to health systems, members and other stakeholders. Sample solutions include member and provider engagement tools, data modeling and analytics, administrative simplification, and provider network solutions.

Kevin Pereau: How are business agreements and partnerships changing the role that health care companies are performing today?

Richard Lungen: That is an important question, since collaboration and cooperation is critical to success. The user of any service, whether a consumer or a provider, doesn't want to have to go to multiple places to accomplish the same or similar task. They value and demand the ease of use of a more streamlined workflow.

There are many examples of that throughout the country. Payers and providers are joining together to create mutually agreeable risk models, and to create co-branded products. That is a critical integration, and we are active in that market trend. Health plans are seeing hospitals and physicians more as partners, and not just names in a provider directory. Providers are becoming more critical partners and stakeholders in health plans.

Kevin Pereau: What new trends have you seen emerge over the last 10 years or so? What do you think the future will be?

Richard Lungen: I would say that many major changes are impacting the industry today. Trends are impacting consumers by enabling more effective use of the health care system, trends are impacting the overall cost of health care, and trends are enabling health plans and medical providers to better collaborate.

However, based on the stakeholder, the priority of how to invest in these trends can differ. The goal is to give consumers new tools to better use the health care system and incentivize providers to deliver effective care.

For consumers, there are new and innovative solutions being provided, whether on their phones or via their employers or insurers that provide tools to transparently explain the cost of a medical event the same way a consumer shops for any other product. Consumers can find different care providers, just like they'd use Yelp to find a particular merchant. Plus, there are tools to better understand the quality of care those providers deliver.

For example, one of our Portfolio Companies enables those things to happen right now, in the consumers' phones, providing important information for the consumer to choose high-value and highly rated providers. It's no different than consumers being steered toward Amazon Prime merchants when shopping online.

Another significant trend is to offer incentives to providers to transition from a fee-for-service model to a value-based arrangement. Increasingly, the mindset is that outcomes are not just billables. There is a substantial trend toward incentivizing with outcomes and creating real financial and risk-sharing partnerships between payers and providers.

Where are things going? I wish I knew the answer. But we could talk about consolidation and integration of care delivery systems. For example, the Aetna/CVS deal will enable 10,000 retail pharmacies to serve millions of Aetna members, using a yet-to-be-created, integrated product.

Another trend, which is only going to gain momentum, is the integration of health plans, providers and other provider stakeholders to increasingly become one vertical stack. What Kaiser has done in various markets is clearly working, and there are examples across the nation with other companies such as Geisinger, UPMC, Intermountain Healthcare and new insurer entrants who are tightly aligning with providers such as Bright Health, Melody, and Clover. It's also worth noting that insurers are acquiring provider groups (like Optum has done) in critical markets, which is creating vertical stacks that allow care to be delivered in a more cost-effective manner.

FUTURE OPPORTUNITIES AND TRENDS AS
CONNECTED CARE EVOLVES

Health care may be going through massive transformation, but it can sometimes seem like we have only taken baby steps toward achieving the changes that will produce fundamental, meaningful results.

Millennials are already driving change and will continue to do so. They are collecting data on themselves. They tend to accept the premise that data should be available both to themselves and to their care providers. They want to connect and share information when it benefits them to do so. And they are impatient with the way things are and eager for the situation to change. Thanks to millennials, the old myth that the consumer doesn't want to get involved with managing their own well-being is beginning to go away. And millennials are teaching that mindset to members of other cohorts.

That change will begin when health care reorients to prevent disease and keep people healthy – not only treat them when something goes wrong. When that change happens, everyone wins.

As this evolves, a wider range of businesses will be drawn into maintaining health and offering preventative, not remedial, care. Nutrition companies, nutritionists, fitness clubs, psychologists and other care providers will migrate from their current position on the sidelines of "mainstream" health care and move toward the center alongside physicians, hospitals and insurance companies.

CHAPTER THREE

THE RISE OF DIGITAL HEALTH

Let's open this chapter by hearing from Beth Andersen, a top health care executive. Few people have a better overview of consumer trends in connected health care. Here are some outtakes from a recent conversation between her and Kevin Pereau.

Kevin Pereau: You have worked for large insurance firms, advised start-ups, and served on the board of directors at Wildflower Health. Is it getting any easier to bring innovative new solutions to market?

Beth Andersen: I think that with the increased focus on meeting consumers where they are with respect to convenience and access, we are seeing an increase in the use of retail clinics, telehealth and online scheduling with primary care physicians. At the same time, there is a greater emphasis on using innovative digital solutions to make alternate types of care more readily available to the consumer. Just as our ways of shopping, and making travel and dinner reservations have changed, we are beginning to see those same concepts apply to health care. The focus is increasingly on convenience, simplifying the consumer experience, providing trusted guidance and recommendations, and personalization.

Kevin Pereau: Later in the book, we will hear from VIM founder Oren Afek. But can I ask, are firms like VIM changing the way we consume health care services? How are consumers changing their behavior? Is this really lowering costs and improving care?

Beth Andersen: VIM is a unique firm in that they do not require downloading an app. They communicate with consumers via text messaging on behalf of the physician. The company has an elegant and simple interface that is focused on consumer and provider behavior.

To have an impact on the cost of care, consumer and provider behavior must be aligned. To close gaps in care and improve health status, you cannot have a misalignment of data about the consumer's conditions or recommendations. VIM provides an Amazon-like experience in terms of selecting your doctor, providing online scheduling so that consumers can make an informed choice and schedule appointments at their convenience. At the same time, the provider is provided with personal information about consumers, so they can have more impactful office visits. The provider and consumer are working from the same data set. VIM also can provide text messages which are powerful in increasing consumer awareness and engagement.

Kevin Pereau: Are large insurers getting better at leveraging technology to manage customer relationships?

Beth Andersen: I believe that consumers trust their doctors more than insurers and as a result, there is only so much that an insurer can do to increase consumer engagement. For certain things like checking claim payments and health savings account balances, the health plan apps are appropriate.

If you are talking about creating behavioral change or using technology as an extension of a consumer/physician relationship and improving care coordination, then I believe that the insurer needs to fade into the background and that developing tools that can support the physician/patient relationship are the future. Examples are making the PCP (primary care provider) selection process much more personalized and like a dating app, providing access to appropriate sites of care and information about the cost of that care. Emergency room, office visit, urgent care and telehealth via online scheduling, and access to telehealth are examples of a personalized, convenient consumer experience that is not just focused on claim transactions.

Kevin Pereau: If Digital Health 1.0 was all about proving people will engage and Digital Health 2.0 was about analyzing all the data generated from the Quantified Self movement, how does Digital Health 3.0, which we describe as "Connected Health," reconnect data and make it an actionable benefit to the consumer?

Beth Andersen: Having that personalized experience based on your own health data that is shared with your provider is powerful. Being able to receive messages from providers if they notice you are not refilling your medications as prescribed, or if you are at the emergency room for a condition that could best be taken care of in an urgent care or doctor office setting are ways to direct consumers to the right care at the right time, in the right setting. This drives affordability, increased satisfaction and over time will change consumer behavior such as not using the ER for non-emergent conditions.

Kevin Pereau: Can digital health help underserved communities?

Beth Andersen: Absolutely, there are many rural areas where there is limited access to care. Telehealth fills that void. Also, people want convenience and if you have an issue at 2:00 A.M., using telehealth is wonderful.

Kevin Pereau: How important is it that we navigate the transition from fee-for-service to value-based care models?

Beth Andersen: In order to moderate the continued increase in the cost of health care, the value-based models are essential. As an industry, we have to focus on increasing the quality of care, access to preventative services and personalized care plans to address individual health status and improve access to appropriate care settings

Kevin Pereau: Any final thoughts on how innovators, providers, insurers and employers can all better work together to lower costs and improve care outcomes?

Beth Andersen: Each stakeholder in the system has an obligation to make health care more affordable. And innovation and technology play an oversized role in that goal. Collaborating to provide tools to consumers could be so powerful to change behaviors, improve the management of chronic conditions, provide access, and make sure people receive care that they need to lead healthier lives at a more affordable price.

THE RISE OF DIGITAL HEALTH

Few people saw digital health coming.

Let's face it. Health care has never been a haven for innovators. Venture capital was lukewarm, and private equity barely had a pulse in health care. All that changed with the introduction of value-based care models. More than $25 billion has been invested, and we now have apps for sustaining engagement, platforms for predictively analyzing, and a way to reconnect those who can help keep us healthy.

It wasn't long ago that statements like these were common:

- "We cannot get around to fighting diseases, we are fighting costs."
- "The system is overloaded, we are overwhelmed, and we cannot get around to tackling the big issues."
- "There are just not enough doctors. If we had more, we would be curing more people."
- "Patients are just not interested in doing what they need to stay healthy."
- "There are so many regulations and laws that our entire industry is paralyzed."
- "The FDA approval process has hopelessly impeded our ability to treat patients with the drugs they need today."

What do those statements have in common? They are part of the blame game, in which every player on the field points fingers at others. You have heard similar opinions. Perhaps you are still hearing them today. Maybe you have even said similar things to people as a way of explaining why you or your organization are finding it difficult to bring about meaningful change.

HOW MEDICAL CONSUMERS CONCEPTUALIZED HEALTH CARE AS RECENTLY AS FOUR DECADES AGO

I know that the observations I am about to make are generalizations, and that generalizations are often not completely true. But in hindsight, medical consumers like you and me expected to be taken care of, and we avoided taking an active enough role in understanding and improving our own wellness. We also never had the tools to do so and nobody seemed to care or advocate for us to make better lifestyle choices.

We would go to a physician for care, and we expected to be told what was going on with our health. Unless we were "health nuts," we rarely thought about nutrition. We exercised during our childhood years, and possibly were athletes when we were in high school and college, but then very few of us continued to exercise. When we saw an adult man or woman running down a street, a sight so common today, we thought, "Who is chasing him?" or, "Should I call the police?"

Bodybuilding gyms were about the only category of athletic facility for adults, though some exercise clubs for women, like Lucille Roberts, were paving the way for the arrival of membership gyms that we see most everywhere today.

There were no apps for health-conscious people to buy back in those days (there were no smartphones). There were only a handful of books like *Let's Cook It Right* by Adele Davis, a pioneering author who posited the idea that vegetables could be made healthier to eat if they were not cooked to wilting. Health food stores cropped up here and there, but they were mostly the province of "nuts" who believed non-mainstream notions that there was a link between nutrition and better health.

The idea that we could educate ourselves and partner with our care providers to create better health for ourselves was still years in the future, certainly not anywhere near the mainstream.

On a more philosophical level, it is even possible that many of us at that time still believed in the notion of fate (or perhaps even fatalism) when it came to health. The idea that we were powerless seemed rational. We heard this statement more often than we hear it today:

"When your number's up, it's up."

Today, that philosophy has loosened its grip on our way of thinking. Most of us know someone whose life, or quality of life, has been positively impacted through the application of medicine and healthy routines. We know people whose lives were saved or extended thanks to modern medicines, surgeries, exercise, diet, and a variety of advanced technologies.

Although the notion of fate has not lost its grip on us entirely, we no longer feel that it is an out-and-out miracle to live five, 10 or even more years longer than our parents did. There really is a brave new world of health care. Most of us are now acutely aware that we are living in it, and we would like to become more involved.

BUT THEN, A FUNNY THING HAPPENED

Gradually, things began to change. Incredibly, it was driven by consumers, not by care providers. Consumers suddenly began to demonstrate that they were not only passively interested, but eager to understand and take control of their own health.

The age of Digital Health 1.0 was dawning. Those "blame game" statements, harder to believe, began to fall away. From the grassroots upward, a new way of considering health care began to emerge that arose from, and centered on, consumers. The dynamic of the industry changed from supply to demand.

THE RISE OF DIGITAL HEALTH FROM AN INNOVATOR'S PERSPECTIVE

Back in 2011, I was raising funds for a new health scoring startup I was helping to launch. The goal was to create a company that would gather and organize data on consumers, and that care providers could then use to provide consumers with better care.

It seemed to my colleagues and to me that the world was waiting for what we were about to offer. But we encountered strong headwinds when we sought funding from the logical sources, like insurance companies. We assumed they would want consumers using solutions like what we were offering. Those consumers stood to benefit the most.

They weren't quite ready to be investors yet, and it wasn't clear to anyone that digital health assets would work. In theory it sounded great, but it wasn't yet in practice.

There was the old blame again, stated in a new way. They had failed to understand that the ground was shifting and that medical consumers were showing they were interested in taking a more proactive role in managing their health. We were offering them an opportunity to not only serve consumers, but to partner with them, thanks to technology, to achieve a better overall state of health.

BIG CHANGES IN AMERICAN HEALTH CARE

In a sense, this book is all about trends. So much is happening, and so quickly, who can make sense of it all?

As an analogy, think of health care as the Los Angeles freeway system- the confusing network of highways that run around and through the city. If you're driving on one of them, you know the scenario. You seem to be moving in one general direction, but a lot is happening. Cars are merging in from on-ramps. Your GPS tells you that events on intersecting highways are going to have a big impact on your commute.

Perhaps you should change your route? Or maybe you are just going to have to deal with your current situation day after day for the next five years, until new highways are built to accommodate the volume?

Who can make sense of it all? The man or woman up there in the news traffic helicopter. He or she, unlike motorists who are in the thick of it, has an overview and can see the overall context of things happening that is invisible to drivers.

While no expert anywhere knows everything that is happening in health care, some very smart people have developed an informed overview of how everything fits together - what is coordinated and what is not.

EXPERT OPINION: DR. MIKE LOVDAL, ADJUNCT PROFESSOR AT THE COLUMBIA BUSINESS SCHOOL

What has changed in American health care, and what will change in the coming years? If anyone knows, it is Mike Lovdal.

Mike obtained his MBA and doctorate at Harvard Business School, then joined its faculty teaching corporate strategy. He later moved to the Sloan School at MIT. Mike then began his consulting career at Oliver Wyman, where over a 35-year career, he undertook projects for clients in consumer products, media and technology. During his final years at Oliver Wyman, Mike concentrated solely on health care serving providers, payers, pharma companies and public health organizations. Today, he serves as an Adjunct Professor at the Columbia Business School co-teaching a course on "Innovative Models in Global Healthcare."

I am delighted and honored that Mike agreed to offer his perspectives and opinions, rooted in deep experience, for this book.

Chances are you realized the following trends were happening. Yet the statistics that Mike offers on the phenomenon are somewhat staggering . . .

In 1960, U.S. expenditures on health care equaled 5% of the GDP. By 2016, that percentage had ballooned to 17.5%, and growth will continue. That means Americans are now spending $3.2 trillion on health care. And according to the Centers for Medicare & Medicaid Services (CMS), here is where it is going today:

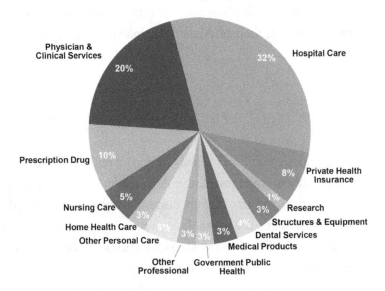

Americans Are Obtaining Health Insurance from Many Sources:

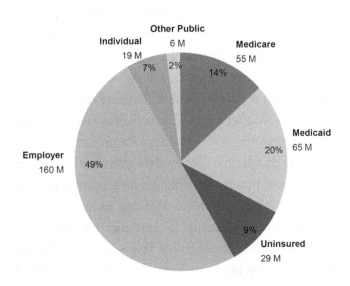

Americans Are Accessing Health Care Information in New Ways

In 2004, only 30% of Americans accessed health information online. By 2015, that percentage had surged to 72%.

But interestingly, that doesn't mean that U.S. health care consumers are relying less on their physicians for information. Back in 2004, 67% of them relied on their physicians for information; in 2015, that had risen to 70%.

SIX MAJOR TRENDS MIKE LOVDAL IS WATCHING

1. State initiatives will trump federal impact on U.S. health care. Even without changes in the Affordable Care Act (ACA), Section 1332 and 1115 waivers currently authorized within the ACA give states considerable leeway in designing their own health care policies. Many states are currently filing for waivers. And every legislative change proposed for the ACA has moved away from a one-size-fits-all federal model to state-based flexibility (e.g., block grants for Medicaid).

2. $2 trillion will migrate to value-based health care. The move from fee-for service (FFS) to fee-for value (FFV) has already started - and is irreversible. CMS has led the way with Medicare initiatives, but the private sector is quickly becoming the main arena for this shift. In many cases, employers are leading the way with innovations such as medical tourism and direct contracting with providers. Payers are also firmly committed to match their cost models with their revenue models. Value-based health care will ultimately become the dominant payment system with major implications for every sector.

3. Health care will become a consumer business. The consumer revolution is underway. High-deductible health plans now offered by most employers have become commonplace,

driving heightened awareness among individuals and families of personal spending, cost options and outcomes. As employers move away from Defined Benefit (DB) health plans to Defined Contribution (DC) health plans, this awareness will grow quickly. And possible changes to Medicaid (e.g., block grants) and Medicare (e.g., vouchers for buying private insurance) will add millions of health care consumers.

Succeeding in this new health care world will require consumer-centric business models for providers and payers with superior expertise in listening and responding to the voice of the health care consumer. Research conducted by Mike while at Oliver Wyman identified ten consumer demands – the first eight for all Americans and the last two for specific segments:

- "I want to understand how to improve my health and live longer and better."
- "I want to manage my health conditions."
- "I want to make informed decisions about health services and spending."
- "I want coordinated, seamless health care."
- "I want personal health information that I control."
- "I want to connect with people and patients like me."
- "I want anytime, anywhere access to convenient care."
- "I want health care tailored to me and my family."
- "I want help with caregiving."
- "I want to live independently."

4. Health care will become an information technology business. We have already seen the proliferation and penetration of Electronic Health Records (EHRs) among acute and ambulatory providers with mixed success. While the major EHR vendors focus on protecting their B2B models and proprietary

systems, consumer health technologies are emerging for each of the needs identified above:

- "I want to understand how to improve my health and live longer and better" will be addressed by wellness apps, tools and biometric wearables.
- "I want to manage my health conditions" will be addressed by condition and disease-monitoring apps and tools.
- "I want to make informed decisions about health services and spending" will be addressed by navigation and transparency apps and tools.
- "I want coordinated, seamless health care" will be addressed by smart care teams with interoperable electronic health records.
- "I want personal health information that I control" will be addressed by personal health records/electronic medical portals.
- "I want to connect with people and patients like me" will be addressed by health crowdsourcing and social media.
- "I want anytime, anywhere access to convenient care" will be addressed by virtual/remote diagnosis and treatment.
- "I want health care tailored to me and my family" will be addressed by genetic testing and precision/personalized medicine.

5. Precision medicine will become commonplace. The science of medicine is exploding with new pharmaceutical, diagnostic and device breakthroughs happening weekly. One of the most impactful examples is the adoption of genetic testing as costs decline ($1,000-2,000 today to $100 sooner than we expect) and the related move to precision medicine (aka personalized medicine). The combination of trend #2, the volume-to-value revolution, and the personalized medicine

trend will set the stage for a major transformation in U.S health care:

- *Today: Reactive, Transactional Individual Sick Care*
- *Future: Preventive, Interventionist Population Health Management*

6. The structure of the health care industry will radically change. The above five trends in combination will drive major changes in the U.S. health care industry with clear winners and losers. Hospitals, the largest single expenditure category, will need to reinvent their businesses and brands as "reform" initiatives and the major trends migrate revenues to new provider models such as national specialty chains and retail clinics. Physicians will continue to consolidate and integrate to protect their independence. Public health organizations will enjoy new appreciation (and hopefully funding) as population health and prevention become the norm. Payers will need to move from simply managing financial risk to becoming health managers and even health care providers. And finally, new models will emerge from private equity and venture capital backed tech start-ups, and unique health initiatives will come from the "frightful five" (Amazon, Google, Apple, Microsoft, Facebook) and their partners (e.g., Amazon with Berkshire Hathaway and Chase).

To Summarize Mike Lovdal's Opinions . . .

Stay tuned for the new U.S. health care reality. We are in the midst of a $3+ trillion "jump ball" where the largest and most personal and political sector of the U.S economy is up for grabs. While the transition will be painful for some, the end result will be a health care system that better controls costs while delivering higher quality, better access, and an enhanced consumer experience.

CONSUMERS ARE EAGER TO ENGAGE

Patients like you and me have always been eager to take an active role in managing our health care, we just didn't know how to do it. And the medical establishment did very little to help us take part in the process. While conventional thinking was that consumers didn't want to engage, the fact is that we couldn't, because didn't have the tools, education and technology we needed to achieve a better state of health.

God bless all those entrepreneurial kids from MIT and Harvard and Stanford who weren't accepting of conventional thinking. They never accepted that consumers wouldn't participate or get involved in maintaining their good health. They simply went out and made great apps and platforms that allowed everyone to better connect.

External innovation is profoundly reshaping the world of health care, creating thousands of apps –the very solutions that people are now downloading daily from Google Play and the Apple iTunes Store.

I came back to the health insurance industry for my next wave of fundraising and I said, "Look we have all been wrong. How can we say that people will not engage or be accountable for their own health when they are using all those apps to do exactly just that?"

At this point, the conversation changed overnight – it just did. Suddenly more insurance companies were inclined to see that customers were becoming more engaged. But would you believe it, entrenched thinking came roaring right back. We heard a wave of new objections that went something like this:

"Okay, people are using those apps now, but they won't be for long. They're a fad and they're going to go away."

I looked straight into the eyes of some of the people who made that second objection and I said, "Really? A little while ago you said that no one would become engaged in managing their own health, and now that they are engaged, you're saying that they won't continue to be for long, because they will lose interest? The technology is just not engaging enough? But do you know what? We seem to be moving in the right direction. We continue to download and use health care apps of all kinds.

Despite early market objections, there was progress. Insurance companies, care providers, and other entities who were part of the health care value chain saw opportunity. I can't claim that I personally wore them down by making wonderful arguments that convinced them. They changed because a kind of tidal wave washed over them – a wave of consumers suddenly demonstrating that they wanted to manage their own health.

Those consumers didn't only demonstrate it by downloading and using apps. They got involved in social networks that brought communities of support directly to them.

SOCIAL SUPPORT ON MANY LEVELS

Social support is a key factor in motivating individuals to take care of themselves. People make the difference. As someone wise once said to me, "You can turn off your Fitbit or ignore a health care app on your phone, but you can't turn off your husband or wife."

Ultimately, those other people are the factors that motivate us and keep us on the right path. If you go to a party and eat two pieces of birthday cake, someone is going to point that out to you. If you're not sleeping well, your spouse or partner is going to ask about that. Those interactions might be irritating, but they are essential, because people who care about us reinforce our positive activities. If you've been exercising every day or if you've lost weight in the last month, somebody is going to comment on that, and they will support your commitment and progress.

Competition can help too. I am now in my 60s, and when I go out and run with men my age, I take inspiration and motivation from them, especially when they're performing better than I am. (I also get a little motivated when I am performing better than they are, but let's not dwell on that.) When I am running in an event or race, the people my age become more motivated by running alongside people who are, say, 20 years younger than we are. We have thoughts of, "Hey, I'm still in the race," or "Those runners aren't so much better than we older people are."

And now that those experiences have migrated to social media, they have become more commonplace, without losing their power to motivate us.

Social media are only a small part of the transformation.

EXPERT OPINION: ORON AFEK, CO-FOUNDER AND CEO OF VIM

If there is one dynamic of health care that is difficult to navigate, much less understand, it is how we engage with consumers at their point of intent.

I'll give you an analogy. When I visit my daughter in the summer in Boston, my trips are always carefully planned. In late July, the Sox have an extended home series that includes the Yankees. I book my flight, my hotel, purchase my tickets for the game and start lining up friends and workmates to attend the games with me. I can book all this based on trying to save money or by staying close to Fenway Park or in Brookline where my daughter lives. Careful planning and advance purchases help me lower my cost and maximize my value. When in health care do you ever have this opportunity?

When I attended HIMSS (the Health Information and Management Systems Society annual symposium) recently in Las Vegas, I wasn't feeling well. HIMSS is the mother of all health care shows so I didn't want to miss the event. I know, I thought, there is a hospital right down the street. I'll just wander into their emergency room, right? Any business traveler can relate. I am out of network, know very little about my surroundings. Why not just go where you know you will be seen? We have all been there. The issue? I have just wandered into the highest cost venue there is to take care of what they would certainly NOT categorize as an emergency. On top of that, I have impeded their ability to see more patients with more critical needs.

What if my phone had pinged me on my way into the hospital to see if it really was an emergency? What if I could book an appointment with somebody in network that was also close by, and I could do that from my phone? What if booking and paying for that appointment with a higher value, less costly provider was all integrated into my benefits and HSA (Health Savings Account)?

MEET VIM

Oron Afek is co-founder and CEO of Vim, a company that partners with medical providers and health systems to provide convenience, cost savings and a better care experience to patients.

A number of other companies are trying to do that, you say? That may be true, but the scope of Oron's vision is larger and different. He is out to revolutionize some of the most fundamental processes that take place in the delivery of health care, as you will learn in these outtakes from a recent conversation with him.

Kevin Pereau: What led you to start Vim, and what is the philosophy behind it?

Oron Afek: I understand why you ask. My vision for Vim is not to compete with other companies in the health care space, but with companies like the Airbnb's and Ubers of the world. Or at least, to apply thinking that is like theirs to health care.

Kevin Pereau: How is that thinking different?

Oron Afek: Well, let's take a quick look at how Google affects consumer behavior. If a consumer is shopping for, say, a new piece of technology like a dashcam or tablet or a Bluetooth speaker and searches for them on Google, he or she usually clicks through on the top search results first and is likely to make a buying choice in that way. That is how the consumer sees the buying process. But behind the scenes, the companies that make those products might see an increase of between 75% and 100% of what they sell.

Why does that happen? It happens because something behavioral is happening in the way people buy things, because of a new way of connecting supply to demand. In a similar way, we are finding ways to make it easy and immediate for top health care providers to connect with consumers.

So we are not marketing or advertising in the traditional sense. We are striving to drive behavioral change.

Kevin Pereau: I'm not sure I understand.

Oron Afek: If you've used Uber or Amazon, you know that they are dealing with very high-frequency demand. If you need to get from Point A to Point B, you use Uber and if you want something delivered the next day or even sooner, you use Amazon Prime. People download their apps and open accounts and use them on a daily basis.

The connection between end users and health care providers is fundamentally different. It is not a high-frequency paradigm. There are thousands and thousands of health apps, but they have low usage rates compared to Uber or Amazon. The people who develop health apps think that if they build them people will come, but it doesn't work that way, even if they use all kinds of incentives.

Health care is generally of a low-frequency nature. You are only going to get sick occasionally, not every day. And when you're sick, you need help now. The secret sauce, if you will, is that we are finding ways to engage patients at the point of their decision-making.

We are willing to educate people and contact them at the moment when they need to make a decision.

Kevin Pereau: Can you give me an example?

Oron Afek: You have a knee problem, you go to see an orthopedic doctor who orders an MRI for you. So there you are in that physician's office, and you need to book an MRI. We are building a system which, at the moment that physician prescribes an MRI, sends you a text message that lets you book it immediately, from a top provider, and at an advantageous cost, with your insurance coverage already approved and ready to go. So we save patients thousands of dollars and provide the best value on that MRI.

Kevin Pereau: How are you able to save consumers money in that way?

Oron Afek: You know, it doesn't really matter if you spend $500 or $5,000 on an MRI, because you are going to have that test performed on exactly the same machine. And guess what? There is no difference in the quality of the MRI. And if you can save $4,000 and you have a deductible, that's going to represent a savings for you. And if we can partner with care givers and offer savings like that, that is compelling I think.

But I'm not talking only about cost savings, but about changing behavior.

Kevin Pereau: How do you identify the times when patients are making decisions, so you can be there at the right time?

Oron Afek: We call that time the "point of intent." We have a team that is dedicated to finding the perfect moment where we can serve consumers with the right value proposition, and to communicate then. It's really about engaging with the right people at the right time and guiding them to the right providers. We are reducing costs without reducing quality at all. We believe in increasing the value of care.

Kevin Pereau: Where did you get the idea of using technology to interact with people at their point of intent?

Oren Afek (laughs): Well, I served in the Israeli Special Forces before I moved to the United States. I was injured and sent to the hospital. I learned that when you are in that position, when you are lying in a bed and you need surgery, you are not the same person you are the rest of the time. You are a different person when you need help, and you make your decisions in a very different way.

Kevin Pereau: What obstacles have you overcome since you started Vim?

Oren Afek: I think it took time to understand where the money is, in the sense of where the value can be found in the world of health care, and what we should do. If you build another urgent care center, maybe yes, that would be needed. But we believe that the greatest value is located in the process of guiding individuals to the right providers, as well as driving providers to do the right things. We think this is a true multibillion-dollar opportunity. Understanding that took time.

Kevin Pereau: Where do you think your company will be in three to five years?

Oren Afek: I think five years from now, we are going to be serving 100 million Americans, and operating in a new kind of place where we can really drive behavioral change on the provider side, and in customers.

Kevin Pereau: What other big trends do you see in health care?

Oren Afek: I know that we often hear that we are seeing a shift to consumerism, but I don't really see that as the biggest trend. I think the real issue is that we are seeing a shift from volume to value.

TOWARD THE QUANTIFIED SELF

The health care industry at some point began to see informed medical consumers not as a problem, but as an opportunity. That was a big shift, a time when the tectonic plates shifted beneath our industry. We saw health care providers achieve a new level of clarity, thanks to emerging data and analytics.

If Digital Health 1.0 was about proving people would engage, then Digital Health 2.0 was when the quantified self movement took off.

From a business perspective, it was a big explosion – a time when investors began to understand that data could be used to not only analyze consumer health care trends, but to make predictive and accurate assumptions about what could be done to achieve better outcomes and better health for many more members of our health populations.

Suddenly, providers became more empowered to help people stay healthy, not just get involved when they became ill. It has become possible today for physicians to know if something has gone danger-ously out of sync, and to intervene through remote patient monitoring and engagement tools. If the data that is being gathered reveals that a patient's blood sugar or cholesterol levels have dramatically changed, or that that patient's blood pressure has suddenly shot up, a connected clinician is now able to contact the consumer to say, "I've noticed that something is happening" instead of waiting until the next appoint-ment and discovering then that a problem has occurred.

THE ROLE OF LIFESTYLE

Back in the 1950s and even the 1960s, the term lifestyle meant some-thing much different from what it means today.

A person's lifestyle had to do with factors that, for the most part, had nothing to do with health. A person who could afford it had a life-style that allowed travel, a nice car, a pleasant home in a secure com-munity, and possibly the leeway to enjoy a round of golf or a game of tennis on the weekend. Somehow, the term was not applied to people who didn't enjoy that kind of life.

Today, the concept of lifestyle has changed. A person's lifestyle, in the common conception, has come to center on factors like these:

- Do you exercise . . . and if so how much?
- Do you engage in healthy behaviors, like avoiding smoking and excessive consumption of alcohol?
- Do you keep your body weight under control?

- Do you do a good job of managing ongoing or chronic health conditions?
- Do you sleep enough, take enough time off from work, and engage in other behaviors that are now commonly recognized as health-promoting?

IMPEDIMENTS TO DIGITAL HEALTH 3.0

So if all those positive changes are taking place, what is keeping us from entering a new, nearly utopian world of wellness? I would say that currently, the biggest challenge could be summarized this way:

We are collecting all kinds data about ourselves, but cannot yet connect it to the people that can help keep us healthy.

In the past, there were many blockages that impeded the flow of data and information. The businesses that developed health care-related products were very far removed from the people who used them. Your primary care physician had data about you, for example, but it tended to remain with him or her.

As data gets more widely disseminated today, we are faced with new challenges and new opportunities. You might be gathering lots of information on a phone app about your diet, for instance, but how can that data be delivered to your physician, nutritionist, and other entities? You are tracking your blood sugar in a glucose meter if you have diabetes, but how can those records be continuously delivered to your endocrinologist or other diabetes care-giver?

The coming age of digital connected care will be driven by savvy players with an appreciation for the importance of data. We will enter an age when insurance companies, hospitals, and other care providers will realize that they will become dramatically more profitable if they learn how to collect and utilize data on their customers. This will make cross-functional acquisitions that extend value chain capabilities beyond the norm. Insurers will focus less on buying other insurance companies and more on retail pharmacy, nutrition, and fitness. The idea being to better integrate everything back into your provider. Integrated offerings will be the rage.

Currently, a wide gap exists between many care providers and the people they serve. But a great delayering is about to happen. In part, that change is already being driven by economic forces and profitability. As insurance companies begin to sell more of their products directly to consumers instead of selling plans to employers, for example, opportunities will arise for more direct communication and data sharing.

WILL OUR DATA BE SAFE?

As medical consumers, we all want to know that the information we're sharing with our primary and other care providers will be secure. Yet when we visit a physician or other care provider and fill out a HIPAA compliance form, we become aware that our data can be shared in some way, with certain entities, under certain circumstances.

What is going on, we want to know? Who will have access to our records? The more that medical consumers can be assured that their data will be secure and safe, the more eager they will be to become fully engaged in digital health care.

DATA AS A CUSHION WHEN EMERGENCIES STRIKE

As I close this chapter, I would like to point out that not every medical event or emergency can be prevented by data. It is unpleasant to think about, but "stuff happens" that apparently comes out of nowhere.

That marathoner you know suddenly has a cardiac event. That ultra-slim yoga aficionado you know suddenly develops high blood pressure. That person you know who avoids sweets and fats and starchy foods suddenly and inexplicably develops type 2 diabetes.

Understanding data can't prevent every health problem. Sometimes, hard-to-explain problems crop up when we least expect them.

But at those times, we discover another good thing about all that data and information we have been collecting. It suddenly becomes useful, if not invaluable, as we work in concert with our caregivers to recover and manage troubling new conditions as they arise. It's another benefit that modern medical consumers enjoy, thanks to data.

CHAPTER FOUR

THE ROAD BACK FROM BROKEN

We have already discussed trends to watch as we march on toward Digital Health 3.0. But trends become real when they are played out in events in the real world – events that serve as mileage markers along the route to a new place.

So in this chapter, let's consider some of those significant markers.

THE AFFORDABLE CARE ACT MOVES INSURERS CLOSER TO A CONSUMER-CENTRIC MODEL OF HEALTH CARE

It would not be entirely inaccurate to say that the incumbents have been reluctant to fully embrace the use of technology much less think of their customers as consumers. Even today, we call them policy-holders and patients. In sum, the mindset has not completely changed for them, and end-users are sometimes still seen as entities who buy policies and whose data appears on P&L statements and actuarial tables.

But it would be unfair to conclude that insurance companies don't care about the wellbeing of their customers

Many of those impediments are built into the systems that insurance companies use to interact with consumers.

Insurers generate a large percentage of their revenue (how much depends on the company) not by selling insurance directly to consumers, but by selling plans to companies through brokers. And once plans have been sold, they are administered in client companies by plan administrators in HR. In larger organizations, there can be even more intermediary people - still more functional links in the long chain that stands between the insurer and the insured.

In other words, insurance companies tend to be steps removed from people like you and me, the individuals who hold policies. And those intermediary entities can really get in the way. More often than not, insurance companies try to gain an understanding of us not by getting to know us as individuals, but by gathering and analyzing data about us that filters up to them.

How many people participate in the plan at Company A, for example? How much money is that plan earning for the insurance company? How many people can be expected to retire or leave companies, and how will those events affect profits? Somehow, the issue of how healthy you or I are as individuals can get submerged, rarely discussed in meetings.

What health-related challenges are we facing in our lives?

Given those impediments in the flow of data, it's very easy for insurance companies to make incorrect assumptions about us as individuals. Many assumed incorrectly that neither you nor I would ever engage to the point where we would share information on who we are, what we do, and how we feel. And if we shared that information, how would insurance companies use it? Is there a place in a spreadsheet where they can slug it in? Usually, there isn't. A kind of "father knows best" mindset has taken hold in the past, based on the mistaken belief that neither you nor I would ever take responsibility for our own health and wellness.

I have been in conversations with insurance company leaders and it has sometimes been difficult to get them to realize that you and I are already actively engaged in monitoring and taking control of our health. Some predict that insurance companies' role will soon be little more than claim-processors. It seems logical and efficient to go on gathering information about us from third-hand sources.

Yet, as expert contributions from insurers to this book show, there is great willingness to find a new way to connect with, and serve, customers.

Ironically (and this isn't a political statement), the Affordable Care Act has served as a catalyst to get insurance companies to start thinking in a new and de-layered way and ask questions like:

"What if there wasn't anyone standing in the middle and we had to make products that were directed and created specifically and entirely for the individual?"

Changes triggered by the law have forced some insurers to question their long-held assumptions. When the ACA was put in place, many insurance companies assumed that most of the people who bought insurance through exchanges would be millennials. But they quickly discovered that in fact, many of the people who were buying insurance through the exchanges were in their late 40s, 50s, 60s. The marketplace dynamics were surprising. Insurers had to confront the fact that they were not making money because they had built products that were not designed to meet the needs of large numbers of their new and potential customers. They suddenly had to interact with people directly. This meant new programs, new services, new ways of thinking. They are also learning that there is no business cost to caring about people.

They have begun to realize that the models they use to run their businesses won't be right for the new world of consumer health.

For insurance companies, the times are surely changing. In brief, insurance companies are starting to realize that to stay relevant, they need to reach out and be in touch us. And how will they make that contact? They will do it through our data.

CASE STUDIES

How are America's largest insurance companies moving into the world of digital health?

AETNA'S STORY

There is probably nobody as fascinating to follow as Aetna among the insurers. They have a visionary CEO who envisions a compelling world of health care in which everyone is connected. Their view of the health care shopping experience is that it should be similar to any other shopping experience we have. From going to the store to reaching for your phone, they are surrounding us with clicks, bricks and mortar to provide us with products, programs, drugs and services.

Their story is one of not being afraid to be in the first wave or of shaking things up. From iTriage to Aetnacare, bespoke consumer markets to their partnership with Apple, they have been a driver at every phase.

While the trend for insurers is to either double down on membership acquisition or integrate provider capabilities, Aetna stood all that on its head when they merged with CVS. Suddenly, a combined firm emerged that can provide you with everything you need, from daily health supplements to prescription meds for those monitoring long term conditions.

KAISER PERMANENTE: A HYBRID INSURER/CARE PROVIDER

Kaiser Permanente is a formidable organization that is uniquely positioned to play a transformative role in the world of health care. It integrates insurance and care, and that is immensely powerful in controlling costs.

They have a tagline that reflects the role they play in the lives of their members:

"Thrive"

Who doesn't want to thrive? And for Kaiser, thriving is more than just a tagline. It has taken steps to help members self-monitor, gather data, and share. They are giving their members Fitbits and other data-gathering devices. Even more important, they are nudging their members toward healthier behaviors and lifestyle choices.

One motivation is to make members cheaper to insure—a philosophy that is built into the way that all insurance companies do business. Yet Kaiser is unique. One reason is that the people who lead it are genuine thought leaders.

Where will Kaiser go in the next few years? How do they continue to innovate when others are embracing Kaiser's model of integrated insurance and care? Some speculate Kaiser will explore acquiring, or partnering with, companies like Jenny Craig or Weight Watchers to expand the footprint of what it offers. I wouldn't be surprised to see new partnerships with retail pharmacies where services like flu shots – and more - will be offered.

Kaiser Permanente will continue to find new approaches to partner with consumers and they will be very exciting to watch.

OPTUM AND UNITED HEALTHCARE: THE BIGGEST INSURERS

United Healthcare is the largest insurance company in the U.S. and is also the most complex. This company insures more U.S. citizens than any other insurer. It has a multibillion-dollar technology division focused on providing digital health solutions to other payers, providers and employers. It also has a large venture fund focused on finding and launching innovative new solutions that solve real-world health care problems. By separating core competencies around distinctly different entities, United has removed barriers to partnering. The results are staggering. There isn't a digital health tool innovator out there getting traction that doesn't land on United's radar.

United has mastered the fine art of smart money in health care. Their investment portfolio has the distinct advantage of being taken to market by the most robust and successful channel in the industry. In addition to money, innovators get accelerated access to customers. United is where you go when you are ready to scale.

Optum is the second "big gorilla" in the world of insurance. And it has become, *de facto*, the industry leader in the use of emerging digital health assets. This company has a cornucopia of solutions targeted at both employers and consumers.

You are probably using some of Optum's products and solutions without even knowing it, because the leaders of this company sell their systems and solutions through several other companies and have become the backbone on which our industry runs. Those companies can offer services that embody state-of-the-art digital technologies and services, and have not had to develop them in-house. And many of them are the technologies that consumers covet and crave.

No doubt about it, Optum is creating a powerful portfolio of tools to engage and involve consumers. When a company purchases a plan from this insurer, that company acquires a powerful portfolio of programs for its employees to use and enjoy. There are walking challenges, health screenings, educational programs, gym memberships, and a lot more.

And like the other insurers I have written about in this chapter, Optum offers powerful online tools that consumers can use to pick their own physicians by specialty, zip code, and other factors. And Optum is breaking ground by helping consumers set up health savings accounts, which help them budget and pay for medical procedures. Let's say, for example, that an employee would like to have foot surgery. He or she can go on their insurer's online platform, learn the cost of that surgery, and see that there is enough money in his or her health savings account to pay for it. That consumer can then have the surgery, perhaps without exhausting the funds in that account. It is all seamless and transparent.

Optum is not afraid to take risks. Their biggest challenge going forward will be consolidating to eliminate redundancy and streamline costs. They realize that they are building a market leader and seem determined to keep their organization in that position.

United Healthcare and Optum are relentlessly focused on better integrating providers with consumers to make their health care experience more seamless. This is what makes them an incredibly exciting company to watch.

United Healthcare and Optum are now actively integrating all the technologies they have either purchased or developed. Redundancy on this scale is not only costly, it confuses the market and commits the company to keep using business models that won't stand the test of time. For these companies, the biggest risk is losing their edge for assessing new digital health opportunities in favor of building out their own solutions. They believe that pursuing best of breed accelerates the speed of innovation.

I look for United/Optum to be long-term innovation drivers in health care, from better integrating payer-provider capabilities to tools for employers. It will be fascinating to see what they do next.

THE STORY OF THE BLUES

In the "barely in the digital health game" category, we have most of the Blue Cross companies. Blue Cross is the most myopic of any of the big players and despite their immense size, it is in perennial catchup mode. What makes it more confusing it that the Blues are comprised of several different companies who make up a mosaic of similar but separate entities. Anthem and Premera Blue Cross are the notable exceptions, and stand out as innovators. Cambia is on the opposite end of the spectrum. It is hard to drive things to scale when you only focus on a specific region.

A management consulting firm would implore most of the Blues to partner with organizations like Optum, which already has a $30 billion head start providing back office health care solutions. They would also advocate finding digital health solutions to purchase or use. There is no point for most of the Blues to spend money trying to organically develop solutions designed catch up with Optum or others who have a lead on the playing field. They should forge alliances with the right partners to differentiate their own core delivery competencies. The Blues have missed the boat. Instead of developing digital platforms organically, their focus should be on making wise use of what others have already developed.

Some of the most remarkable innovations in health care are taking place where employees, employers and health care providers meet.

EXPERT OPINION: RAJIV KUMAR, PRESIDENT AND CHIEF MEDICAL OFFICER, VIRGIN PULSE

Rajiv Kumar, M.D., is the President and Chief Medical Officer of Virgin Pulse. Virgin Pulse, which is part of Sir Richard Branson's Virgin Group, designs technology and cultivates good lifestyle habits for employee groups. The company's stated mission is to ". . . drive more meaningful habits, for more employees, than anyone else. And we're proud to say we're changing lives."

As of mid-2018, Virgin Pulse supports 1.9 million members in 17 languages across 185 countries.

Rajiv Kumar is a seasoned businessperson who is committed to building global health. He has learned unique and useful lessons as he has helped Virgin Pulse achieve a preeminent position among health care companies. Let's hear some of the insights he offered in a recent conversation with me.

"What obstacles have we had to overcome?"

What obstacles have we had to overcome? One of the biggest in the health care space is that it's very noisy. There is so much innovation happening, there are so many startups, so many people turning all their attention to trying to solve big problems and impact a lot of people and improve the quality of their lives.

So one of the biggest obstacles we had to overcome, and which I think any health care company must overcome, is to ask, "How do we become the signal in that noise?" That is another way of asking whether it is possible to stand out from the pack.

Another issue is that there a lot of bad actors in the field. There is a lot of proverbial snake oil being sold and as a result, buyers are skeptical about whether or not the solutions they are hearing can make the change and produce the results that their developers are claiming. So buyers and investors have to evaluate CEOs, companies, and vendors that are coming at them. That is really, really critical.

"We are not going to be another company that is doing the same old thing . . ."

From our earliest days, we were determined not to be another company that was doing the same old thing. We didn't want to get lost in the crowd. And it really is a crowd. By our last count, there are more than 900 wellness companies in the U.S. alone.

We decided that we would turn conventional wisdom on its head, wisdom that holds that although company wellness programs are critically important, it is impossible to get employees to fully take part.

Is it really true that people don't want to take part in their companies' employee wellness programs? Well, we are not so sure. Employees might not want to take a health assessment or get their blood drawn. And some of them might not want to use a company's outside fitness facility or take part in company fitness challenges and programs.

Prevailing wisdom held that it was pointless to spend money or pay people to try to get them to change their behavior. Paying people to engage? Most people agree that really wasn't the answer.

And so we asked, what is the opposite of a financial incentive? We decided to tap into intrinsic motivation, which is the opposite approach. We went into the market hard with that point of view. We used financial incentives for *vendors* really as our foil, so we started to sell our company in an unconventional, out-of-the-box kind of way. And that's the approach we took for differentiating ourselves, for getting people to pay attention to what we had to say.

At a certain point, we began to realize that it is not an either/or kind of situation, because people do engage in different ways, both through financial and social incentives. The answer lies somewhere in between. We have to leverage both, to be open to all the different kinds of motivational styles that people have. So we started to evolve.

"We had to get to scale as a company . . ."

Another obstacle that we had to overcome, and which I think many entrepreneurs have to confront in the space, is that to get to scale as a company, they have to sell to large enterprises, whether they be large employers, large health plans, or large hospital systems.

But there's kind of a chicken and egg problem. Those big entities, specifically in the health care space, are very risk-averse. They don't want to be the first to try something new. They require loads and loads of evidence and of course, they're very focused on security and compliance. Those are some of the fundamental challenges in the space.

How do you sell to an enterprise when you're not at scale yet? How do you get the scale? That's one of the things we have really tried to overcome.

The solution evolved around finding some innovators inside of large enterprises, and to convince them to take a risk while containing any potential collateral damage. Sometimes, allowing them the ability to wall off the segment of the population, or carve out a piece of what we can do, maybe in a pilot format. That allowed us to prove our value without asking companies to put everything that they were building at risk.

That helped us start to work with large enterprises, and to prove ourselves And I think that's one of the important things that everybody needs to know about the space.

"The payers are finally waking up . . . "

The payers are finally waking up to the fact that they cannot build every kind of relationship themselves. Even if they could, they're not necessarily going to be well accepted by the market. And so payers are definitely opening up to the idea of partnering with best-of-breed, point solutions by which solutions reach employers and their employees.

I have not seen a ton of examples of early-stage startups achieving wild success by attempting to hitch their wagons to a channel partner early on. That doesn't seem to work as a market strategy. The reason is that while your products are evolving, you need to stay as close as possible to the end-user or the end customer. By going through a channel, you can create a wall between yourself and the end-users and lose touch with what they are actually looking for, how they are experiencing your product, what they're missing when they use it, and what the value proposition truly is. So you can miss out on involving your product and your company in a way that's going to allow you to capitalize on the opportunity you have.

And we've certainly stumbled over that a number of times, thinking that partnering with a large health plan or large broker was going to be our ticket. That being said, now that we are north of $100 million in revenue, we're actively looking for distribution partnerships. We've got the product, we've got the evidence, we know it works, it's rock-solid. Now it's simply about getting as many people as possible. And people realize, it makes much more sense to go to a channel through a channel partnership. Knowing when the right time is to do that is really important for startups.

"You've got to raise a bunch of money . . ."

Billions of dollars are flowing into digital health. I think it's very hard as a self-funded or bootstrap company to go head-to-head against big companies that are raising $50 or $100 million. That's something everybody needs to be aware of. If you want to play in this space, you've got to raise a bunch of money to be truly effective.

Now, are there companies that bootstrap it and get started successfully? Absolutely. But to get to scale and to carry out a brand vision in the health care space requires a lot of capital.

"It's very important to identify who the thought leaders are . . ."

I think it's very important to identify who the thought leaders are in the industry, and who the gatekeepers are. Yet they aren't necessarily the same people who are funding your business, and it's likely they won't be.

I think people make the mistake of assuming that the people who are the investors will be the same people who will open the doors to the kingdom, who know what the real solutions are.

So you have to forge alliances with the people in the industry who can actually help you. And they are usually not the same people who are funding you.

"We are going through a sea change . . ."

Is the patient/end user experience improving? Absolutely. We are going through a sea change in the evolution of the products that are available to end users. We are seeing what I call B to B to C products and solutions, like Virgin Pulse, where we are one business selling to another business, and that business is selling to consumers.

Solutions like that, I would say, have reached parity with B to C solutions. There are companies out there that are delivering B to B solutions that are as good as, if not better than, what a consumer can buy on his or her own. And customers want that kind of quality experience when they are interacting with businesses. That is what it takes to establish credibility.

The Virgin Pulse app is one example. Eighty percent of all interactions between end users and us happen via that mobile application, versus our web application. Our app is actually the highest rated mobile application in the App Store. It's more highly rated than many consumer applications.

To be truthful, I think we only got there in the past year or so. Other companies are catching up to that as well. It took us about a decade to get there, but we learned, and we improved, and we certainly borrowed lots of ideas from the consumer side.

I think that is a key to success, because consumers won't tolerate solutions from their providers or their health plans that are subpar, compared to what they can get elsewhere. If a health plan offers subpar solutions, that plan gets cut out of the equation and isn't able to affect any change in the way that it wants to. I think that's a really important point.

"I foresee a lot of consolidation . . ."

I foresee that a great deal of consolidation is going to happen in the space. It's very fragmented as we reach this kind of first decade, and go from 2.0 to 3.0, in the terminology that you, Kevin, use. I think there's going to be a shakeout, a healthy consolidation throughout the industry, not that we want only a small handful of players controlling any space. We want a lot of innovation and a lot of competition. But at the same time, I think there are a lot of solutions out there today that are subpar, and that waste time and waste cycles. So is a shakeout coming? It is going to happen for sure.

"It's very important not just to have a point of view . . ."

I would also add something that we have learned about leadership in this industry. It's very important not just to have a point of view, but to publish that point of view and share it widely, and openly discuss and debate it. That's what we consider thought leadership.

In our case, we engage experts and conduct surveys, and put up data to show the industry what is happening, where it is evolving. That is why good companies take us seriously. If you are in the space, companies want to know in-depth information about your products and solutions.

At Virgin Pulse, we will continue to invest heavily in thought leadership through our science advisory board, through our events, through our whitepapers and more. Leading with thought leadership is critical. You have to invest in building up evidence early and often. It sometimes takes two or three years before things take off. But having hard, scientific, credible evidence is critically important, preferably in the form of published research.

"You have to pick one market to go after . . ."

One mistake we made early on was that we failed to pick one market to go after. A lot of companies make that mistake.

It's very tempting to think that you've got one product that is immediately going to serve multiple markets. It is tempting to think that you have this wellness platform that you are going to be able to sell to employers, to providers, to employees. You believe that same product will work for everybody. We learned quickly that each buyer has a totally different value proposition, with a different set of pain points. It is hard to bend and twist one product into so many different archetypes that will meet different market needs.

I would say a lot of early-stage companies are trying to sell into three or four different markets, and that almost never works. Our industry is littered with the carcasses of companies that have tried to do that. So that would be another big piece of advice for innovators in health care: focus, focus and get your niche.

EXPERT OPINION: GRANT VERSTANDIG, FOUNDER OF RALLY HEALTH AND CHIEF DIGITAL OFFICER, UNITEDHEALTH GROUP

Every industry has a few key players - individuals who have established their roles as true innovators. In the world of digital health, one such innovator is Grant Verstandig.

Grant's story to date is unique. A *Fortune* article (October 2016), "How a 27-Year-Old College Dropout Is Simplifying Health Care," gives an overview. *Fortune* reports that when Grant was a freshman lacrosse player at Brown University, a knee injury and multiple surgeries forced him to quit the sport. While navigating the health care system and recovering, Grant had an epiphany of sorts and realized that his life's mission was to simplify the way people navigate the health care system and manage their health and well-being. He dropped out of college soon after that and in 2010 started what would become Rally Health. Today, Rally has joined forces with UnitedHealth Group and employs more than 1,000 people. In 2018, *Business Insider* included Grant on its list of 30 leaders under 40 in health tech who are using technology to shape the future of medicine.

RALLY HEALTH BELIEVES IN SIMPLICITY

Rally's mission, posted on the company's website, reflects Grant's seminal belief that health care needs to be simplified:

"To be effective, health care needs to be simpler. Our goal is to break down complex topics and ideas into actions, empowering consumers to take control of their health before facing crisis or injury. And should

they become injured, quickly finding the right doctor and getting up-front costs should be easy.

"As a digital health experience, we are focused on members' overall health. Personalized daily goals, recommendations, and rewards are key for prevention, while cost transparency and access to information are crucial when it comes to getting care."

The goal, the website copy states, is *"To put health in the hands of the individual. That energy permeates everything we do."*

We are pleased to include this edited conversation between Grant Verstandig and Kevin Pereau, in which Grant reflects on where health care has been in the past, and where simplifying it will take it in the years to come.

GRANT VERSTANDIG ON FORCES THAT ARE DRIVING CHANGE IN HEALTH CARE TODAY . . .

I think you're right about the stages of health care evolution that you describe: Health 1.0, Health 2.0 and Health 3.0.[1] That is sort of the macro landscape. Something I would add is that socioeconomic and sociodemographic changes have been a major driver of change, and a driver of engagement for Rally, too.

Specifically, every day there are about 10,000 people who age into the Medicare system, and when they do, they tend to utilize more care and spend more money, not less. At the other end of the spectrum, we see that Millennials are not consuming health care services in the way

---

[1] See Chapter One. Digital Health 1.0 was the time when the popular notion was dispelled that consumers wouldn't engage in their health management; Digital Health 2.0 was the time of the Quantified Self movement and the rise of predictive analytics; Digital Health 3.0, happening now, is the time when we are extending the health care value chain to include nutrition, exercise, and mental health

many expected they would. They tend not to have a traditional physician relationship, and so they basically don't utilize care . . . until they *really do*. And when they do, it's an enormous jump that tends to spike around things like complex neonatal care. As you know, each day in the neonatal intensive care unit can cost $10,000 or more.

Further, across the board - from Millennials to seniors - we are seeing a tremendous increase in the rate of obesity and its associated chronic illnesses such as diabetes and heart disease. That population used to be about 30% of the people in the United States. The number is expected to be as high as 50% of the population in any given state in 2030.

The interesting opportunity for Rally Health is that there are now large segments of the population that are increasingly engaging in the health system and could benefit from using Rally, our consumer digital health platform, to help them manage all the complex elements involved in their health and health care, from understanding their benefits, to meeting their health and wellness goals, to finding care providers and pricing care.

ON THE EVOLUTION OF DIGITAL HEALTH . . .

Today's digital health solutions have come a long way to deliver real value in people's lives. In the beginning, that wasn't necessarily the case. When thinking about digital health, people would say, "Okay, I just go online and do all these digital things . . . but does that make the health care experience easier to navigate, less expensive, or more effective? Does it make it so that the care I get for my mother is easier, or for my son, or for myself? Does it simplify things or lower costs?"

Today in digital health, we can answer "yes" to all of these questions. As an example of what digital health today can do, at Rally, we can tell our members who are prediabetic and interested in losing weight, "If you lose more than 5% of your BMI [Body Mass Index –

ed.], we will lower your health insurance premiums." And, "If you use our coaches, you will save even more." We make it easy for them to do these things. They can simply click a button and have a live, digital coaching session. More than 90% of all the coaching sessions we do now are digital; just a few years ago, 90% were telephonic. And we are finding that that is exactly what people want.

ON THE VALUE OF SIMPLICITY . . .

Simplicity is where we think the magic really happens in digital health. When we were first developing our Rally platform, former Apple CEO John Sculley, who was on our board, would tell me all the time that it is what you leave out of products—not what you put in them—that gives the consumer an "AHA" experience.

We have to remember that however simple we make health care seem to a consumer, it's extremely complex behind the scenes. And that's the point of digital health - to take complex issues and processes and make them simple for the consumer who wants to be empowered to discover information and resources that are specifically relevant to them.

At Rally, we aim to meet consumers where they are in the health care system and simplify whatever they are doing or experiencing at that time, no matter *what* they are doing. If they are searching for a physician, we can meet them there and help them find one. If they are trying to figure out what the best preventive cancer screening would be for them, based on their data, we can help there, too. And each of these processes feeds others.

ON REDUCING COSTS
AND IMPROVING OUTCOMES . . .

If you are about to interact with the health care system, Rally can help you make the most informed and best choices for *you* based upon factors such as your specific medical condition, the physicians or facilities located near you, their quality ratings, and pricing.

For example, imagine the experience of a first-time mother whose child has an ear infection. She could use Rally to quickly discover her options for care. She might see that a visit to the ER could cost up to $10,000, and that a visit to the urgent care center would cost less than going to the ER, but could still be pretty expensive. She would also see the option of scheduling a virtual visit and could have the antibiotics to fight her child's infection delivered right to her home in less than 24 hours. She would see that this choice is significantly less expensive. The point is that by using Rally, she would learn about her choices, including lower-priced options that could meet her needs just as well as higher-priced options.

A big part of what we are doing now is taking data and using it to make personalized recommendations for our members. Similar to how Amazon Prime makes recommendations based on what consumers have bought in the past, Rally can say, "You are seeing a primary care physician, but did you know there is a higher quality premium care physician in your plan who is closer to you than the person you are seeing now and could offer a 10-18% lower total cost of care?"

If you look at the overall value of the digital health support offered by Rally for consumers, the cost outcomes become really compelling. For example, UnitedHealth Group reported that when a consumer selects a premium designated physician, they reduce their costs of health care by more than $300 per care episode. In 2017, United saw over $100 million in medical costs saved for customers through our Rally Care solution.

ON WHAT'S COMING UP AT RALLY HEALTH . . .

We're launching a digital-first Medicare Advantage product, replacing the usual paper-first approach to communication. Our goal is to lower the cost of care by several percentage points for Medicare Advantage members by bringing to market a personalized, digital-first, mobile-first product that helps them make better informed health care decisions.

We have already found that for our Medicare members, there is a 1000% increase in engagement when they are offered a cost-saving incentive, compared to about a 200% increase in the rate of engagement for people who are between the ages of 41 and 55.

Medicare consumers in particular want to understand what they are spending, and how they can save money. Many of them are retired, health care is expensive, and they are concerned about the cost of care. We expect that our members will not only be surprised, but delighted to see how Rally can help them get better and more convenient care for less money.

We're a mission-driven company, and this is an example of Rally trying to tackle one of the biggest challenges in health care: managing and serving the needs of the rapidly growing Medicare population. The idea is to apply the same technology and experience we use to serve our current member base to also serve those who are a part of something as large and complex as Medicare.

ON SUPPORTING HEALTHY BEHAVIOR . . .

At the very foundation of Rally is our commitment to support healthy behavior. Our Web and mobile solutions help people get healthier by supporting them in making simple changes in their everyday routines, setting goals, and tracking their progress, while earning rewards along the way. We offer programs for losing weight, reducing stress, quitting smoking, and much more, that are all tailored to the individual and give them the support they need.

One good example, which I mentioned earlier, is our digital weight loss product Real Appeal, which focuses on addressing obesity and its associated chronic diseases. Its digital coaching sessions can have a major impact on helping people set goals and understand the importance of eating better, exercising more, and staying on track to safely and steadily lose weight.

Simplicity comes into the picture here, too. Right now, if you Google "portion size" or "perfect plate," you are going to get hundreds of pieces of information, some of which offer conflicting advice. Our digital weight loss product concentrates on the literacy piece around nutrition and helps people understand how to plan a plate, for example, and what kinds of foods are recommended for them.

Contributing to Real Appeal's adoption is the fact that we have said we are not going to charge the consumer a penny for it. If you are a Rally member who is eligible for the program, based on your BMI, you can use Real Appeal. By taking the cost of our program away from the consumer and tying the program to better outcomes for insurance companies and employers, we have created a positive proposition for all parties.

In just a few years, our Real Appeal program has helped hundreds of thousands of people lose a combined total of nearly two million pounds.

ON WHAT DIGITAL HEALTH WILL LOOK LIKE IN FIVE YEARS . . .

I will say the most important thing by a factor of 10 is personalization, which very simply means that how I use my health data is very different than how you use yours. My diet, exercise, metabolism, etc – it's all different from yours. Each of us will be able to take different approaches to our health and health care depending on our own unique health profile, and we will rely on digital technologies to help us.

Personalization is what has made and will continue to make tech wildly more successful and stickier than any other industry. Look at Netflix: It understands what you liked to watch in the past and from that, suggests what you might like to watch in the future, and it's available on demand whenever and wherever you want it. Digital health is taking health care in this direction.

It's important to the future of digital health that we continue doubling down on delivering personalization. Not only because of what you and what I want and need from our personal health care can be so different, but also because there are things that are different about our local environments. Health care is exceedingly local, and how we take that personalization and link it both to each person's health and their local environment will be key in the future of digital health.

SUMMING UP THE COMING AGE OF HEALTH CARE . . .

Today, our Rally consumer digital health platform already has more than 20 million registered users. When you look at the potential impact that digital health can have from a public health perspective, the potential that numbers like this represent is truly mind-boggling.

For example, we could leverage the reach of a digital health platform like Rally to undertake a public health campaign to encourage everyone to get a flu shot. In fact, we could address any number of core issues similar to the way that we took on smoking over the last decade. Smoking went from being cool to barely acceptable, and then to "you go outside." The next example might either be something not-to-do, like smoking, or it might be something to-do, like getting a flu shot.

At the end of the day, health care is a series of very small decisions that people make in the course of any day, and how we help people enhance their decision-making power, and engage them and empower them through digital health to make the right decisions about their health, that's the future of health care. It's as simple as that.

EXPERT OPINION: MARIO SCHLOSSER, FOUNDER AND CEO, OSCAR

Mario Schlosser earned his Master's Degree in Electrical Engineering from the University of Hannover in Germany in 2002. During that time, he was also a Visiting Scholar in Computer Science at Stanford University. He then earned his MBA at Harvard Business School, and went on to launch several technology companies that included Vostu. Mario also became involved in technology investments, an interest that he pursued at Bridgewater Associates (where he was Senior Investment Associate) and other companies.

Then in 2012, Mario founded Oscar, an innovative health insurance start-up that allowed him to combine his passions for technology and healthcare. Oscar is a consumer-focused health insurer, unlike any other, that is dedicated to using technology to humanize the experience of customer care. Among other benefits, Oscar offers its members quick access to telehealth consultations with plan physicians, a dedicated healthcare concierge, a proprietary and full-featured health management app for all members, and a health clinic in Brooklyn that offers primary care along with wellness activities like yoga.

Mario is recognized as a visionary in the world of healthcare, both for what he has accomplished and for his informed view of where health care is going in the future.

We are pleased to include some of Mario's insights, which we quote from a recent conversation he had with Kevin Pereau.

Kevin Pereau: At the time you started Oscar, all of us in health care were experiencing what seemed like gale-force headwinds. Although investors thought that healthcare was a good place to be, many held back from getting involved, thinking that consumers would never become fully engaged in managing their own care. Insurers were caught up in old ways of doing business and often seemed to be several levels removed from the people who used their products. Hospitals and caregivers were operating on a model in which they generated income only when people became ill, and there was not much business interest or investment in keeping people well.

Yet from a different perspective, 2012 was a wonderful time to be involved in health care, because everything was broken, right? So there were a lot of opportunities that were created by all the problems.

Mario: Definitely, there were problems in both the good and the bad sense, right? Exactly right.

When we started Oscar, we talked about the fact that health care was way too complicated and way too costly as well, and that those two things were obviously related. And that they created opportunities too.

My wife was pregnant at the time. We couldn't figure out the bills, couldn't figure what was going to happen next in her care, and didn't feel like anybody was out there watching over what we ought to be doing next – not our physicians, not our insurers, in essence, no one.

We thought that there was a real need for a different kind of entity in the world of health care, a company or health care system that would have your back. The vision behind Oscar was as simple as that. So we envisioned a company that would be watching what was going on with you, one that would recommend and provide necessary services on your behalf, with the goal of helping you achieve better outcomes and lead a healthier life.

And we thought that an insurance company was in a very powerful position to do all those things in theory, because an insurance company is positioned to have reams of data, to get all the bills for all the services that consumers encounter, and one that can put all that information into the system. That's a very powerful place to be from the standpoint of impact.

Kevin Pereau: Do you think of Oscar as a health care insurance company or as a health care technology company?

Mario: I think we are 50% a technology company and 50% an insurance company. But being a tech company, and not just a health insurance company, means we have a certain way of thinking. Technology enables us to put the members' experience at the center. We try to design systems around the question of, "What will the member experience look like if we were to build it from the ground up?" rather than, "What can technology do?"

But one reason that I think we're more of a tech company is that tech companies are accustomed to taking risks. And I think the art and the science of risk-taking is going to have to be learned by pretty much everybody in health care. We have seen that already. More and more hospitals are dealing with risk. So are more device-makers, physicians.

So the big transition in the U.S. health care system in the next few years is going to be that currently companies and care providers are secretly happy when costs go up, and that has to change. You cannot, in my view, have such a system. You have to orient it more around enabling the players in the system to own the member experience end to end, and for the member's entire life.

That is not the case now. Currently many insurers expect that their customers will be with them for two or three years on average. At Oscar, we'd love our members to be with us for two or three decades. We want to be with them for their entire lives. We want to help them know how to achieve much better outcomes, and we want to be there with them for the long term. That kind of orientation, I think, represents a very big change.

Kevin Pereau: That will be a huge disruption. As a result of that change in the way insurers think and act, do you think that the roles of doctors, hospitals and other care providers will somehow follow suit and change too and orient around the customer experience?

Mario Schlosser: I do. As we know, the way health care is currently being reimbursed is generally on the model of sick care, not interventional care. But overall, more entities are starting to realize the importance of investing in preventative care and wellness.

That change has to happen, has to change, for sure. I have seen big longitudinal studies of health care that show that something like 35 or 45 percent of a person's entire life actually revolves around health care. So I think we are going to have to start looking in a really serious way at the issue of how care is delivered.

At Oscar, we are starting to look at ourselves almost as though we are a subscription service, not an insurer. Members pay a monthly fee, and we reinvest most of the money we make in developing better outcomes for them.

If we can really look at our company in that way and assume we can eventually keep our members for a couple of decades, that changes things. As a member, you invest that fee in us, and we invest in on your behalf in new care and wellness offerings, and you get a lot more value that helps you stay well and avoid becoming ill in the first place.

Kevin Pereau: How do end-users – your members- perceive that kind of change?

Mario Schlosser: One thing we clearly have noticed over the last few years of building this out is that when we can get members engaged in one aspect of our product experience, they start to use more of our products for other parts of managing their health care as well.

When we first launched, for example, we began to pay members a dollar a day for walking a certain number of steps. One of the first things we noticed was that after we started doing that, participation in that program also drove up our telemedicine utilization.

It makes sense that when people start to use one aspect of our care, they discover and start to use other offerings too. There is a big button on our app that lets members talk to a doctor immediately, for no additional fee. People discover and use that, and that leads them to make more discoveries and use them too. And the longer we can keep you as a member, the longer we can train you and help you discover and engage more. And the more data we have, the more we can anticipate what is going to happen and what you are going to want. And the more we can anticipate and offer services accordingly, the more we can engage our members.

Kevin Pereau: As you look back, can you describe any missteps that were made as companies moved into applying digital solutions to health care? Where did health care companies get it wrong in the earlier generations of the marketplace? Did we expect more young people and millennials to be buying services from their phones, for example?

Mario Schlosser: In the insurance market, it took a few years for things to shake out, before insurers could get to the place where we were pricing things just right. The government and the regulatory bodies simply didn't get a lot of things right in the beginning, and didn't get on top of trends quickly enough. But I think those were a couple of difficult years that we all had to go through.

The other thing that happened, I think, was that the health insurance companies didn't know how to rethink the service model. They weren't prepared to really connect with their members. Their first efforts with new technology came via interactions with a bunch of middlemen - the brokers, the consultants, and so on. So it was a much more raw market, one they didn't know how to deal with.

There were shakeouts and companies closed or remained stalled, and that put more pressure on the ones that remained, to try to make this work, like us. It took a while, but I do think we are beginning to see some real utilization on the part of health care consumers.

I think another thing that the market got wrong was focusing too much on premiums and too little on the total cost of care. You know, it is cheaper for our members to have free telemedicine included in their plans, and to have access to urgent care. It is much better for them because we know exactly which drugs they are taking, which care providers they are seeing. And we can track them and talk to them.

Using that data, we can connect them to our own internal systems, to the right doctor at the right time, for example. And when we can connect our members to the right physicians at the right time using our data, that keeps costs down and dramatically improves the quality of care at the same time.

There is a kind of learning curve problem to address, because too many people today end up buying the premium without necessarily knowing what kind of care they are going to get. I think that's one big problem, but I think that is going to shake out over time, as people come to understand that the experience matters. A plan is not good because of its cost, but because of the quality of care it delivers. We, like other companies, have to find a better way to highlight that in our communications to the marketplace. One obstacle is finding ways to communicate that fact to consumers who are in company health plans, because the HR departments, you know, kind of get in the way.

Kevin Pereau: I sense that with some large insurance providers. You could bring them the shiniest new object in the whole world, the greatest new idea to improve health and the customer experience. But if they don't know how to get what you are offering out to their customer base, they are not interested. And at the same time, they're sitting on a mountain of data about their customers that they could be using.

Mario Schlosser: It's a huge question. It helps explain why one of the biggest uncertainties we faced from the ground up was whether we should be an insurance company. We wondered whether consumers would ever trust an insurance company to help them get the best advice on their care. End-users, based on their prior experiences, often don't think that any insurance company would ever take on that role.

But for us, the answer to that question is becoming very clear nowadays. We have a higher engagement level with our members than any other insurance company out there has. One reason is that our members have tried us out, and have seen that what we are offering works. For example, about 40% of Oscar members have used us to find a primary care physician for the first time. Similar numbers of our customers have used us to find a PCP, a specialist, and so on. And they are using our app to talk to a physician or book an appointment.

Part of the high acceptance rate is shaped by our interface. When you call us, a human representative always picks up the phone and talks with you. When you use our app to connect with your care team, you see the names and pictures of the six people who are on it. One of them is always a nurse. And approximately 40% of our members use our app or Concierge service to route them to the kind of care they need, and connect to a real, living person who has access to their data. And when those members see Oscar recommended physicians and other caregivers, the cost of care is on average 10% lower than what consumers are paying elsewhere for the same kind of care.

From the very beginning, we provide a unified experience – we explain to you and remind you and connect with you. And that is far different from the patchwork solution that consumers experience with other insurance companies.

You have had the typical experience of connecting with a physician who doesn't have your medical records, who doesn't know what medications you are taking, and so on. At Oscar, we understand that when you call us or chat with us, there is a 70% likelihood that you are going to have a physician visit within the next four weeks. So in that contact with us, we have a nurse come on the call who will say, "Let me help you find the physician you need . . . let me make an appointment for you . . . let me go over your medical history and medications." Then the doctor you are going to see gets that information before your visit. If you have been working with other insurers, you will be surprised how this works. Other insurers are perhaps too busy to direct you to the care you need, or to lay the groundwork that leads to a doctor visit that is better, more efficient, and less costly.

Kevin Pereau: You're making me want to become and Oscar customer. Plus, that kind of preliminary experience takes guesswork out of it. For somebody with diabetes or hypertension, what a big difference that can make.

Do you have an investment fund, so that if some of those best-of-breed innovators come in with technology you can use, you can partner with them, and let them come in and be part of what you are doing?

Mario Schlosser: We don't have a fund. It's something we have been thinking about. When it comes to partnership, we often prefer to develop technologies and systems ourselves. Some of those decisions can be complicated.

As an analogy, let me mention that when developing the iPhone, Apple decided not to use an email app that had been developed by third parties. Apple knew they had to build the most important apps and the most important parts of the consumer experience themselves. And we tend to think in exactly the same way. Because otherwise, we realized, we couldn't really control the way data is used throughout all of our system. In general, I think the important consumer experiences should be built, tested and refined internally. We are quicker at building and testing and refining if we do it internally. Telemedicine, for example is one functionality that is very important to us.

Also, when we build various modules internally, it makes it easier to share data between them. When we build them, it is easier to share.

Kevin Pereau: This all seems simple and well conceptualized, but it has profound implications for how consumers will experience and manage their care.

Mario Schlosser: I am excited to see what is going to happen in the next few years.

CHAPTER FIVE

THE REAL VALUE OF CONNECTING

Growing up in Vermont, my brothers and I were always moving. Winter brought skating and sledding and skiing. Warm weather brought baseball games, games of catch. My brothers had their sports, and I had mine.

It was a physical life. Even the daily business of getting where we needed to go involved a lot of physical activity, in the form of walking or, in warm weather, bicycling. Our family didn't jump in the car every time we needed to go to the market. We had to walk to get to school every day, no matter the weather.

We ate whole foods, though we didn't call them that at the time. Our family had a garden, and not a "hobby" garden. It was big enough to supply food for our family for at least two seasons out of the year, and really more. No pesticides, no GMO's, of course. Working with my dad in the garden was how I contributed to our family's food supply. My mom would take the tomatoes and other vegetables and can and bottle them in Ball jars. She would make red sauce from tomatoes and we would make it last as far into the winter as we possibly could. She'd pickle things too. We were growing and making our own food.

My brothers and my dad liked to hunt, and that contributed meat to the family's supply. I didn't like to hunt, but I did like to go fishing. My dad had me lined up with a neighbor, a Native American man, who seemed to know all the good places to put in a line, and we would go out together and I would come home with fish to contribute.

Every year, a grass-fed steer was our main source of meat for the winter. We would buy one and basically took it down to a meat cutter in town, who cut it all up. Like most Vermonters, we had a big freezer. Ours was down in the dirt basement.

THEN VS. NOW

So that was basically what life was like back in 1965. How have things changed for young people today?

In the "similar" column, I would argue that young people, both back then and today, are (hopefully) healthy, with nothing but energy and curiosity and enthusiasm about learning about life.

Youth is a time of self-discovery, a time when an evolving person has not become set in his or her ways. The world is a wonderful, big, broad place waiting to be explored. And youth is all about exploration. Some young people are athletes, some musicians, some poets, some budding scientists who, from a young age, poke around and study the world and discover that butterflies come from caterpillars, and many other realities too.

Part of self-discovery at a young age is all about finding and connecting with other people who share interests, enthusiasms, and hopes for the future. You love to play chess or baseball or the piano, and you find people with similar enthusiasms, and you foster one another's growth.

In those respects, young people today might not be too different from people who were growing up 40 or 50 years ago. But even though they generally have remained eager, hopeful, optimistic, and eager to connect with others who share their enthusiasms, a lot has changed in the world. In no particular order, I would point to these changes:

- Technology has made it easier for young people to learn, but less knowledge seems to be passed directly from person to person, or from an older person to a younger one, except in classroom settings. I don't know how many young people today, for example, have an older fishing mentor like the one I had, who showed me where to fish, how to fish, and more.
- Social connections today are often via social media, or using text messaging and apps which, of course, did not exist four or five decades ago. People still connect person-to-person, of

course, but today there is also a technologically enhanced way for people to connect and get to know each other.

- Many young people today are driven to excel in academics, athletics, entrepreneurship and other activities. Young people today seem to have less down time to discover what resonates with their interests and skills, because they are always too busy studying, taking SAT prep classes, going to team practices, and following more rigid routines. I am not saying that is necessarily bad, just that it is different. When I was a kid, I had the time for hobbies, and those hobbies allowed me to discover and pursue interests that were not necessarily taught in school. And then we come to the way people eat today.

ENTERING THE AGE OF PROCESSED FOOD

I suppose that in the years when I was growing up, some food was mass-produced. To be sure, there were some manufactured foods to be found in the then-small freezer cases in markets. There were frozen vegetables and berries, "TV Dinners," and other manufactured products. There were also breakfast cereals that had been manufactured far away, often, from the locales where they were for sale. But 50 years ago, the age of manufactured foods hadn't yet arrived in full force. It was the age before mass production hit the food industry.

Talk all you want about how myopic and profit-driven the health care industry can sometimes be. But, if there's a true villain in harming our health and wellbeing, it would have to be the food industry.

At a certain point along the way, food manufacturers began processing and packaging foods that were jacked with sugars and preservatives. They contain fat, chemicals and GMOs and other ingredients that are genuinely bad for people, but processed sugar is the biggest culprit. And over the last 50 or 60 years, food manufacturers have begun to put sugar into nearly everything they make.

It is amazing. If you pick up any product at the supermarket and read the ingredients, you will discover that sugar is almost always the second or third biggest ingredient . . . in just about everything! And that is true when you are shopping at Whole Foods, Kroger, Safeway, I don't care where.

And here's the obscene thing. Much like the tobacco industry, the people who run food companies know that what they are doing is bad for you. They absolutely do. They've spent billions of dollars researching and understanding all aspects of sugar – why it gets people addicted to foods, how it contributes to weight gain and diabetes, how it shortens life expectancies, and more.

The food industry knows that a very big crisis is coming. It will look like the crisis that hit the tobacco industry. Behind closed doors, there are plenty of discussions about what is coming and what to do about it. Governments around the world will soon be waving their bony fingers and saying, "You know what you have been doing has been bad. You've been creating diabetics, hypertensives, people with heart disease, basically killing people, but you went right on and you did it anyway." It is already happening today.

The restaurant industry has been just as bad. Forty or 50 years ago, most restaurants were serving meals that were made from fresh, locally sourced ingredients. That is no longer the case. Many moderately priced and even upscale restaurants typically serve foods that have been prepared and frozen in factories that are located some distance from where the meal is served. Those foods generally contain lots of fats and sugar – sugar is everywhere.

Fast food restaurants typically serve food that is awful for people but easiest on their budget – and they have been doing it for decades. The American way of selling unhealthy fast food is now being marketed in other countries. You walk down the street of any beautiful European city that you've read about since you were a child and the first thing you realize is, fast food places have popped up on some of the most cherished and coveted streets worldwide. And much of the food they serve is horrible for people, and more people are discovering and eating it. Sadly, the rest of the world is catching up with the USA when it comes to nutrition and obesity.

Yes, fast food restaurants have made progress toward serving healthier foods. But improvement has been slow. When McDonalds began to serve salads, for example, they were being augmented with large quantities of fat so that they would appeal to the typical McDonalds customer. Those salads contained bacon, cheese, but few fresh vegetables. And they were slathered with high-fat "creamy" packaged dressings that were full of preservatives and other chemical additives.

Over the last 50-odd years, an interesting trend has evolved as good nutrition has migrated from the countercultural edge of society toward the center. In a sense, things have come full circle, as people are finding ways to get back to eating the kind of foods obtained from local sources.

Will large numbers of today's consumers begin to grow their own food? Not likely. But more of the food they consume will resemble the healthier foods that they should be eating. That revolution is already taking place.

THE YOUNG AND HEALTHY MINDSET

Young and healthy people are constantly exercising, watching what they eat and connecting on social media about their exercise and health activities. And something rather amazing has happened – their elders are behaving in many of the same ways. As baby boomers age, they are returning to what they did when they were younger.

Today, more older people are sticking to healthy patterns of eating and exercising well into their later years.

They are preserving more youthful attitudes and enjoying much more robust health as a result. People are all about social interactions. They surround themselves with like-minded people. They enjoy sharing their experiences with others and value getting input from others.

A few years ago, I decided I would walk to the train station every morning, instead of driving or taking an Uber. It is three miles. When I first started doing it, I would be all sweaty by the time I got there — so much so that I had to get a locker at a gym I use near the station, so I could shower and change my clothes before taking the train to work.

But before long, I didn't have to shower. The walk was taking only half the time it took me initially. The whole notion of gaming began to kick in and I started to think, "how can I do even better?" I decided after a while that I would also walk home from the station, which meant I was walking six miles a day. I continue to challenge myself to find more reasons to walk places.

When I realized that I was walking something like 10,000 steps a day, I set a goal to increase that. Instead of parking close to the door of a store I was shopping at, I would park as far away as I could. When going to the upper floors in buildings, I began to take the stairs instead of the elevator. Because of those changes, I got closer to walking 15,000 steps a day, which is about 1.5 times the goal for most people. To get there, I really had to go out of my way a bit, but now I am challenging myself even more. If I realize on a Thursday that I have already walked my weekly goal of 12 miles, I don't call it quits. I see the opportunity to increase my challenge.

I am not telling you this to impress you. I am telling you to illustrate that although I am now in my sixties, I am still exercising and focused on eating well.

THE CHANGING WORLD OF ELDER LIVING

When I was growing up, many of my friends' nuclear families included one grandfather or grandmother — most often it was a grandmother who, widowed because most men died in their 60s back then, was living with one of her children's families. This is a lot different from what often happens today.

In some ways, having an elder in the home was good. He or she could live in a family setting, sometimes help with parenting and housekeeping routines, and not go off to live in an impersonal setting. On other levels, I wonder whether having an elderly family member living at home was always best for them. If an elder's health declined, how many of their children were equipped to change their diapers, monitor and administer medications, catheters or provide the kind of care required?

Growing up in Vermont, I used to deliver milk from a local dairy to what we called old folks' homes. I was struck, time and time again, by how badly some older people were treated in them. I saw horrible things. Dining room attendants who rapped residents' knuckles with wooden spoons when they did not finish their meals. Caregivers who belittled them for having bowel movements too soon, requiring a diaper change at an inopportune time. Elders were being punished for being old.

Today, our more fortunate elders have opportunities to live in wonderful communities, sometimes when they are only age 55. Many of these communities have pools, golf courses, social halls, and active programs for exercising and attending cultural events. Many offer pharmacies and programs of health and medical care that are graduated to meet residents' needs as they grow older, often offering assisted living facilities and even hospice care when and if it becomes necessary. In short, these places are radically different from those old folks' homes I saw when I was a kid, and in some ways, I suspect, provide radically better care than could be available in a home setting.

More and more, facilities like those are gathering and maintaining data on residents and sharing it with care providers and families. If you have a parent who is living in one of them and he or she is missing meals or has just stopped attending exercise classes, you are going to know about it.

NEW ENVIRONMENTS

Elders, as noted, are moving into new kinds of caregiving, care-monitoring communities. At the same time, baby boomers are entering their later years, moving from the residences they occupied during their prime earning and parenting years, into communities and residences that are digitally connected in remarkable new ways. And if we look down the age ladder a level or two, we already see members of Generation X and Generation Y surrounding themselves with digital devices of all kinds for interacting socially, monitoring health and exercise, and just plain living in our new digital age. We are all living in a time when people are sharpening their minds and disciplining themselves to be able to think through complex issues in a linear fashion.

When you boil it down, you realize that the smartphone is the impetus and the enabler for much of the change that is taking place. I did not come to that realization entirely on my own; it is a point made by Dr. Eric Topol, who says that the smartphone, more than any other device, has emancipated medical consumers by bringing them new sources of information, connecting them to each other, communicating their medical data to platforms, and much more.

EXPERT OPINION: AMIR DAN RUBIN, PRESIDENT AND CEO OF ONE MEDICAL GROUP TALKS ABOUT THE NEW CONSUMERISM IN DIGITAL HEALTH CARE

What could be more disruptive and frustrating than a trip to the doctor? For many people, not much.

You have to book an appointment well ahead of time. The doctor's office is located far from your office or home. Getting there is a hassle. Once you arrive and sign in, you wait, wait, wait in a stiff chair in a packed waiting room. A TV is showing programs that you would never think of watching. You update your medical information on paper forms that are stuck on a clipboard. The pen is running out of ink and when you are done, you discover that the magazine selection is half of what it was the last time you were there.

And when they finally call your name and the doctor sees you, your time together lasts only 10 minutes. And for much of that time, he or she is pounding on a laptop.

It can all add up to an exasperating experience. What would happen if your visit was designed and oriented around you, not around your caregiver's business?

That is happening, and fast, at One Medical Group.

When you first visit OneMedical.com or start using the One Medical App, you are probably going to think, "Where has this company been all my life?" And with good reason. One Medical dramatically simplifies everything you need to see a physician - fast - and then to get back to your busy life. One Medical currently offers its services at central locations in Boston, Chicago, Los Angeles, New York, the San Francisco Bay area, Seattle and Washington, D.C., with more locations to come.

Ninety-five percent of One Medical appointments start on time. Appointments average a substantial 30 minutes, at a time when most appointments with physicians today typically last 10 minutes. Blood tests can be administered on-site, eliminating the hassle of traveling to have tests done. If you need a referral to a specialist, that can happen in a day or two in most cases. Virtually all medical plans are accepted. One Medical automates the process of filing claims and does everything it can. Plus, One Medical is now beginning to offer services for anxiety and depression. And membership costs only $15/month.

If it all looks like the doctor's office of tomorrow, it could well be.

We spoke with Amir Dan Rubin, One Medical's President and CEO, to ask him about his company and where it is going. Amir is easily one of the most forward-thinking entrepreneurs in health care today.

AMIR RUBIN ON THE PHILOSOPHY OF ONE MEDICAL

"One Medical is designed to transform the delivery of health care through a differentiated model of primary care that improves the patient experience of getting high-quality medical care. We are out to transform health care, and we are doing it through a member-based, technology-enabled primary care system.

"The members who enroll with us are more than just patients. They are people we proactively engage with on health and care issues. We deliver an outstanding experience and have achieved a 90% positive Net Promoter Score. At the same time, we have demonstrated to employers and payers that we can reduce emergency room visits and hospitalizations by up to one-third. A lot of that is because we offer such great service, 24/7 digitally and virtually, and then same-day or next-day appointments in our offices."

REDUCING INCONVENIENCE TO NEGLIGIBLE LEVELS

"Our approach provides a frictionless access to care. Our offices are located where people work, shop and live, not in medical complexes. We offer digital access - patients can have a video visit with a caregiver that lasts three to five minutes. We are for people who really value their time, on their own terms."

BECOMING A ONE MEDICAL PATIENT

"We fit right within the health care system, and we disrupt health care from within. You go online, you sign up as a member. There's a very low monthly membership fee, and we bill insurance on the backend. So we fit within the current system, but we kind of fit on top of it.

"Also, increasingly, a number of employers are sponsoring the membership fee for their employees. Some are very large companies like Google, GE, NBC, PricewaterhouseCoopers and Uber, who offer us as a benefit to their employees. But then we also have individual consumers who go in and sign up online.

"Our model fits within new, broader themes of consumerism and value-based care. We have a different operating model. We have our own technology and build our own software. We have our own electronic health tracker, our own digital platform, our own visit-management system, our own system for managing workflow. All our providers are our own providers, they're all on a salary, not employed in a fee-for-service role.

"Our technology helps move work across the enterprise. We have people who are online reviewing member questions and queries, 24/7. Members get return calls within an hour or two and receive return emails within several minutes."

HOW ONE MEDICAL USES PATIENT DATA

"I should say, we are a longitudinal primary care company. We have longitudinal data on hundreds of thousands of members. And we are using their information to deliver proactive engagement - sending them reminders, giving them advice, and using data to help guide our operations internally.

It's one thing to have a lot of data, but it's another to use it at the right time, and for it to be accessible without 1,000 clicks. We apply a combination of human−centered design thinking and process thinking, used by outstanding people.

"The goal of our entire system is to encourage care givers to spend more time with the members. That is achieved by a combination of things that I believe are so impactful - not just data, not just process, not just people, not just incentives. We are combining all of those things."

BIG TRENDS AMIR RUBIN SEES

"The health care system is difficult to navigate. It's a complex system with a complex interface. Confronted with that, people want frictionless access. We are striving to be like your iPhone screen – a very simple interface. We help our members navigate the system in several different areas that include primary care, referrals, benefits, and advocacy.

"Another major issue is delivering value. You know, the U.S. health care system is swelling up to 18% of the GDP. Meanwhile, expenditures on primary care are less than 6 percent of the overall spend in health care. But we can impact 10 times that spending level. Spending very little, we can deliver great value, particularly since our system has great providers, paid on the salary model, promising evidence–based medicine, and backed up by the data and systems to coordinate patient care."

EXPERT OPINION: HENRY LOUBET, CEO, BOHEMIA HEALTH, ON THE ROLE OF THE HEALTH CARE CONSULTANT/BROKER

If you have ever signed up for a medical plan through a company or a union, you know it is a fairly simple process. You sit down with someone who explains what your plan will cost and what it provides, and you sign up. There might be a few additional steps, like selecting a primary care physician, but you're pretty much finished interacting with your plan until you go get a checkup or submit a claim.

But on that day, you begin to form an impression of whether your new plan is a good one. If it covers a large portion of your medical and pharmacy expenses, you probably feel pretty good that you have landed in a good plan. If it bundles in some meaningful extras at low or no additional cost (a health club membership or a Fitbit, for example), you start to think it is a *really* good plan. And if you have recently developed type 2 diabetes and the plan offers a no-cost membership to a great diabetes coaching and monitoring service like Omada Health, you think you have been lucky enough to have signed up for a *great* plan. Your employer is going the extra mile to provide meaningful extras that will improve your health and make you feel really good about your coverage.

That is what you and I and other consumers see from our vantage point. Rarely do we stop to consider questions like, "Where did this plan come from . . . how did my employer pick it and customize it for me and my colleagues?" Most end-users assume that some kind of salespeople sold the plan to our employers, and that's how we ended up with it.

That could be true. But in most cases, the plan became available thanks to the efforts of a health care consultant/broker. These are individuals and companies that serve as intermediaries between the companies where we work and insurance companies and other providers. As I write this book, about 95% all companies and organizations in America obtain their insurance and health plans not from salespeople *per se*, but through the services of broker-consultants.

A thoughtful, resourceful and caring consultant/broker can make a major contribution to improving the quality of care that end users receive.

Bohemia Health is a major health care consultancy/brokerage located in San Francisco. Bohemia's clients include EHealth & Medicare. com, Stanford University CERC, Omada Health, Burnham Benefits, Pinnacle Brokers Insurance Solutions, Alecto Health System, MCOL, EmpirRx, Brainguard, Antelope Valley Cancer Center and Delta TPA. Prior to starting Bohemia Health, Henry was Western Region CEO of United Healthcare (1996-2000) where he was responsible for 1.6 million members across 10 western states from California to Colorado.

How do quality plans evolve?

"It starts with talking to companies and other entities that need plans," Henry explains. "The idea is to develop solutions - be they insurance plans, technologies like Fitbits, and specialized services that enhance the value proposition for the organization that will offer the plan."

The decisions that result from that process can affect the health and quality of life for a very large group of health care consumers. That was the case when Bohemia Health worked with the Monterey County Schools in California, a buying consortium made up of nearly 20 school districts that employ thousands of individuals in communities that include Carmel and Salinas. After working with the Monterey Schools to assess their needs and desires, Henry and his team decided to partner with Anthem. And because Bohemia Health stays ahead of the curve on developments in digital health, it was able to offer modern extras that included companies that included Teladoc and Castlight.

"Employee benefits are the second highest cost that companies incur," Henry states, "second only to payroll. First salaries, second, employee benefit costs. That means medical, which is probably the biggest of those services and expenditures."

The attitudes that are present, even in some cutting-edge digital health companies, can be another obstacle that prevents end users from getting access to their products. "Digital health companies don't understand or utilize the power that distribution has over the way plans are implemented," Henry says. "Some larger companies want to go direct to employers, and they can sometimes do that. But by and large, companies rely on distribution partners."

"Consumers have needs," Henry summarizes. "Insurance companies and digital health companies have solutions. Broker/consultants are kind of sitting in the middle. Sometimes younger digital health companies don't understand the decision-making chain. When I sit down with them, I am sometimes surprised by what I call limited knowledge and limited relationships.

"Also many broker/consultants aren't adventuresome enough. They will say, `let the Anthems or Aetnas do it,' and just try to sell products instead of finding new solutions for the marketplace.

"They should be advocating for the new solutions," Henry says, "not saying, `That's not my role.'"

It is gratifying to learn that consultants like Henry Loubet, like so many other individuals who occupy influential positions in the world of health care, are conscious that they are not just "making deals," but moving the needle and improving the quality of care that large numbers of Americans receive.

LOOKING AHEAD

And what do we foresee when we look ahead and try to imagine the future of health care and what will mark the milestones?

Soon, just about everything will be equipped with a sensor, from the car you drive to your home, to your bathroom scale to your blood pressure monitor. Data will not only be collected, but monitored, measured and analyzed.

One of the biggest challenges is, who owns our data? Who has access to it? Who gets notifications and why? Blockchains in health care will result in a true tectonic shift in who owns our health care data.

Health is the new wealth. We will still define success by how nice our house is or the zip code we live in. But going forward, health is going to become increasingly synonymous with social status. Health *will* be a currency of its own. You cannot necessarily buy health, but you will know how to earn it, and you earn it every day of your life.

You have to earn your health and you do that by being more active, better connected.

CHAPTER SIX

DATA-GATHERING DEVICES AND HOW WE USE THEM

Consumers use digital devices for a variety of reasons. On the medical side, we find implantable and worn devices like defibrillators, pacemakers, and a growing variety of blood glucose monitors and insulin pumps. There are also devices that measure and record blood pressure, body temperature, body weight and body mass. Then in the fitness area, we have all kinds of devices and apps that keep track of our exercise patterns, heart rates, workout routines, and more.

In the past, these devices have generally fallen into three categories:

- Caregiver-prescribed medical devices that require FDA approval. They are prescribed for you when you are in a physician's office or in the hospital, because you have a condition or disease that should be monitored. You go home with them. They are typically used to keep tabs on issues that are serious, or which at the minimum require monitoring. Glucose monitors are probably the most common devices in this category.
- Medical devices that medical consumers decide to buy on their own. Perhaps you get a diagnosis of high blood pressure from your caregiver, who suggests that you get a blood pressure cuff and begin to use it for self-monitoring at home. Or you decide on your own to buy one, just to keep tabs on the situation surrounding your blood pressure. So, with no prescription required, you drop by a pharmacy and buy one.
- Fitness monitors that are informative, motivational, and fun to own. They help you keep track of your exercise routines, and help you stick to your fitness program. They've been around for years, they keep getting better – and now they are migrating onto smartphones.

But categories are merging, blurring and increasing in both effectiveness and number.

CHANGING EXPECTATIONS

A few years ago, many patients thought they were keeping tabs on their health if they were getting on the bathroom scale twice a day, or tracking meals in a diary, or keeping records on their fitness routines.

Today, consumers are monitoring their own health and wellbeing in more ways. When you are outside your house, these systems keep records on how much you walk, how many steps you take, whether you are exercising at a high altitude, and more. Inside your house, they can track nearly everything you do, such as how much you are moving around the house, how many times you go up or down the stairs, how many hours you sleep (both at night and when taking naps during the daylight hours), when you watch television and use your computer, the temperature of your home, how many times you open the refrigerator door, your pulse, your breathing rate, what your galvanic skin response is – and on it goes.

One example is a health suite app I use that has an internal gyroscope that knows whether I am sitting or standing. It reminds me when I have been sitting too long, and sends me a message that it is time for me to stand up and move around – or remain standing while I work. I have it programmed to ping me twice a day to remind me to do a breathing exercise, and to lead me in that exercise.

Not long ago, people were asking, "Why would anybody want to track all that, or to have reminders all day long about what I should be doing?" But that attitude is changing quickly. Many people, probably led by the aging Baby Boomer generation, seem to think that it is just fine – and even desirable – to have data collected so extensively and constantly, and to use electronic devices that nudge them to stand up, sit down, walk, or engage in mindfulness exercises.

Objections are disappearing, and I predict will soon be remnants of the past. The explosion of available apps and devices to track exercise and diet is one of the driving forces behind the change.

AN EXPLOSION IN FITNESS MONITORING DEVICES

Several years ago, it seemed curious to see someone wearing a strapped-on heart monitor while running. Today, many people wear them, and nobody pays much attention. And more and more exercise monitors are becoming permanent parts of clothing – you put on your Under-Armour shirt, it contains monitors that collect data on your run or workout and sends all the info to your smartphone. Not long ago, you had to put the on-chip monitoring devices into little pockets in your exercise clothes. Now, more electronic chips have been sewn by clothing manufacturers into the garments we wear when we workout. Soap hasn't always been friendly to those microchips, but that should change soon, because we are entering an age of flexible, encapsulated microcircuit boards that will happily survive trips through the washing machine – and even the dryer – without failing.

When my friends ask me to recommend a digital fitness monitoring device, I sometimes suggest that they should read product reviews on *Consumer Reports*, and I am not being facetious. If they take my suggestion, they will find reviews of a larger selection of fitness monitors than they thought possible. Fitness monitors are available from many makers, including:

- Adidas
- Charge
- Fitbit
- Flex
- Fossil
- Garmin

- Huawei
- Jawbone
- Kua
- Life Trak
- Mio
- Misfit
- Moov
- Polar
- Samsung
- TomTom
- Under Armour
- Withings
- Xiaomi

The prices start about $20 and run up to several hundred dollars. Which device is best? I don't know for certain, but why not just buy one and start to use it, if you're interested? The price point is just so small.

And then there are all the apps you can install on your phone or tablet – literally thousands of them, too many to list. The chances to keep tabs on your exercise are increasing daily.

I invite you to meet two brilliant young men who occupy a position at the forefront of applying digital assets to fitness and exercise.

EXPERT OPINIONS: TRAVIS SHANNON, VICE PRESIDENT OF INFORMATION AT LEISURE SPORTS HOSPITALITY AND BILL DANIELS, ELITE PERSONAL TRAINER

Between trips to the hospital, people are looking for ways to stay healthy. That helps explain why no health care segment has grown faster than the fitness industry. It provides us with the facilities, expertise and resources we need to get and stay healthy.

It is not uncommon today to see gyms focusing on nutrition, relaxation, chiropractic and full spa services. If we see the doctor when we are broken, then the gym is where we spend our time when we are fit. Savvy gyms are adding devices for tracking and connecting us via social networks to better influence our activities and improve our fitness results.

The challenge for fitness clubs will be figuring out how to better connect their member data with their members' primary care providers and insurers when they are sick. Everyone in the world of fitness wants to be an influencer, and the industry is just warming up to the idea that relationships are longitudinal and not just based on our impractical moment of need.

And a bigger trend is at work. While the merger of CVS and Aetna is today's industry blockbuster news event, mergers and cross-functional acquisitions are happening in the world of exercise and nutrition too. If payers and providers can be more profitable with healthier members and patients, then why not add resources that show them exactly how to do that?

Travis Shannon is Vice President of Information at Leisure Sports Hospitality in Pleasanton, California, an industry leader in designing and managing upscale exercise and health clubs. Bill Daniels is a certified elite personal trainer who serves clients at Renaissance Club Sport, a top exercise club in Walnut Creek, California.

Both these men are leaders in the world of health and exercise and bring uniquely informed perspectives about how the use of digital data is impacting the world of exercise.

"I've been in technology for a long time in various facets, prior to joining Leisure Sports," Travis says. Before joining Leisure Sports, he worked in software development in the financial and consulting industries, as well as in technology commercialization at Los Alamos National Laboratories. He holds a Bachelor of Science degree in both Computer Science and Biochemistry from Hobart College, and in 2005 graduated cum laude with an MBA from Babson College.

What is a digital specialist like Travis doing at a company that designs, owns, and manages high-end fitness facilities and resorts? The fact that he is there reveals a great deal about the current trend to integrate digital technology with fitness.

"I came to the company to build the IT department and to align technology and business," Travis says. "The fitness industry has been, and does continue to be in some aspects, woefully behind in technology. But you're starting to see the industry entering a modern age. So it's been fun to be here to watch that. We are a company that has been around for well over 35 years now. We were built as a company that managed health clubs. Today, our consumer-facing brands are Club Sport, Renaissance Club Sport, and The Studio."

Club Sport is the company's original brand, now a high-end presence in the fitness world – a kind of "fitness resort," as Travis puts it, that offers not only exercise equipment, personal training services and classes, but also extras like massage, beauty salons and even on-site sports bars. Renaissance Club Sport clubs are like Club Sport facilities but have a Marriott Renaissance Hotel combined with the property; Renaissance Club Sports operates both the hotels and the spa/fitness facilities – a "nice blending of two business models," in Travis's words. The Studio clubs are smaller, stand-alone exercise facilities that are currently expanding into new locations.

What trends does Travis see, from his perspective?

THE IMPACT OF WEARABLE TECHNOLOGY

Travis says that the widespread adoption of wearable devices has "bled into" the world of fitness facilities. A few years ago, Leisure Sports began to work with MyZone, a company well known for selling a chest strap heart rate monitor.

Within Leisure Sports' exercise clubs, MyZone technology is used in a variety of ways. A client's heart rate and other data can be displayed on the screen of the piece of exercise equipment he or she is using and tracked remotely when he or she is on the cardio or exercise floor. If customers are in a spin class, a large screen that the class sees displays tiles for all cyclers with data about their heart rates, current speed, distance covered, and more information. Data collected from MyZone devices will not only monitor activity on-site but upload it to the MyZone platform when the wearer is riding a road bike or running outside of the facility. More and more, data is being collected from wherever and whenever clients exercise.

Initially, Club Sport's reason for integrating more fully with MyZone was to increase the amount of time that members spend using exercise facilities. "We wouldn't want to give everybody a piece of technology and say, `you don't need us any more,'" Travis explains.

"Regarding wearables, what you have seen is the growth of communities and platforms that increase facility use and make sure that people know how to stay on track. But the data and the technology, in and of themselves, are not necessarily the driving force that gets people to exercise."

He believes that technology motivates because of its ability to let clients share information and join communities where they have conversations about how the data relates and matters to their health and wellbeing.

"That's the area where we are focusing," Travis says. "The technology helps, and it helps us reach beyond our four walls and keep people involved."

TECHNOLOGY AS A MOTIVATOR

Travis believes that technology is more than about "counting steps." It is something that can motivate people to get off the couch and exercise.

Trainers and exercise professionals can monitor individual clients' exercise records and data. In that spin class, for example, an instructor can say, "Okay, I want to see everybody in the yellow range." But such feedback is varied by individual - not reported on a one-size-fits-all scale but adjusted according to different people's levels of fitness.

Technology also provides opportunities for customers to interact. Club Sports runs special events, like Member Challenges, that use the MyZone app as a platform for tracking people's progress in a virtual competition. People can join actual teams, virtual teams, or compete as individuals over a certain amount of time as they pursue their individual exercise routines. Whoever ends up with the most effort points wins.

Travis views such gamification as a big aspect of how technology can be used to build commitment to exercise. "People love it, and it becomes something they want to participate in," he says. "And it is especially effective when you combine a social aspect, with people forming teams. But it is not the technology in and of itself. People often ask, `how accurate is the individual data that is collected,' but that is a secondary concern. If you can get individuals to become engaged and active, and to start to take their personal data into account when they make decisions about their exercise and health, then you have won."

A WIDENING NET

Health clubs are beginning to use technology to help members receive care and counseling from a wider network of care providers that can include nutritionists, chiropractors, and other care providers. Says Travis, "We see the potential in partnering in those areas."

DATA INTERACTIONS WITH INSURANCE COMPANIES

Travis believes that more insurance companies are becoming interested in partnering and interfacing with fitness companies. "For a long time, insurance companies have subsidized health club memberships for their members," he says, "because they have known that individuals who exercise are better people to insure. But we are now seeing an increasing interest."

He also notes that accountability and trackability are on the upswing. In the past, some individuals would use the club memberships that were provided by their insurers to get into health clubs, then go sit by the pool or go to a restaurant on site. New technologies that track members' data on how they exercise could usher in a new level of cooperation between health clubs and insurers.

"I think you are going to see insurance companies implement trackable activity requirements for their customers who are using health clubs," Travis says. "But maybe not. Maybe engagement is enough, and is what insurers should be looking for, without diving in deeper."

Another trend to watch will be a growing number of connected exercise facilities that companies are installing on site, for the use of their own employees. "There will be aggregate data collected about how many people are using those company centers," says Travis, "possibly leading to wider collection and use of individual health and fitness data."

PERMISSIONS ISSUES

Are people sensitive about how their data is being collected and used? Travis feels that at this point, his company is more concerned with check-in and usage data, and uses it as a way to improve operations, expand into new markets, and grow. "There is some basic stuff that we track," he says. "But our policy has always been never to provide customer data to any outside sources. But I think that more and more, insurance companies are accepting self-reported data from their customers."

Where do individuals obtain data that they can share with their insurance providers, if they opt to do so? They get it from the apps and technology they are using to monitor their activity. The self-service portals in Club Sport, for example, allow clients to collect information on their own exercise routines, levels of fitness, and other data that they can opt to share.

"I think we are now on the cusp of really digitizing and streamlining the process. That is something that the fitness industry is behind right now, while hospitality and health have jumped up. But we are starting to gather a fuller history of who our members are, what their goals are, and what their challenges are. We are just on the cusp of really seamless interaction. I think that all fitness companies are really right there too."

"The use of technology to enhance the clients' experience and to enhance the understanding of the benefits that they receive for being members, is where we are seeing the most benefits, and where we are most excited."

CHANGING EXPECTATIONS

Travis believes that in the next few years, customers will make their membership decisions based not on comparisons between one health club and "that other one down the road."

"They will be won over when they can say, `I interact with this club in the same way I interact with Amazon.com, or with Nordstrom, or with my bank. I can walk in, or I can communicate and interact using my phone. There is a nice kiosk or other simple point of contact. And when I come in, they have my information, so we can pick up where we left off and I don't have to provide it all again or cycle back around."

In short, Travis notes, "There is a lot of opportunity."

Meet Bill Daniels . . .

Bill Daniels, a certified elite personal trainer at Renaissance Club Sport, a top exercise club located in Walnut Creek, California, has been named Best Personal Trainer in the San Francisco Bay area by *Diablo Magazine*. He especially likes working with clients who are recovering from previous injuries that have limited their ability to exercise.

"The way insurance companies tend to work is that they want to get you to a certain point in your recovery," Bill states, "and then they release you as soon as they can. Those people are generally not back to 100%. I like to work with those people to get them there. It could be something as common as tennis elbow, but I also screen people for potential future injuries and do corrective exercises before they get hurt." Bill also incorporates some neurology in what he does. He has a lot to say about the role that digital information is playing in the lives of his clients.

DATA COLLECTION

"It's funny," Bill says, "because I really didn't realize how much I use data until I knew I was going to talk with you about your book."

In his club, trainers use MyZone, which started out as a heart rate monitor. Clients install the app on their phones and use it to track their exercise and nutrition. "On my phone," Bill says, "I can see where all my clients are with that. I can monitor how often and how long they're exercising, at what intensity. I can see how much they have eaten, and what they have eaten.

"Sometimes I have to touch base with them to verify what they have eaten and what they have done. For example, if I look at the MyZone app and see that they worked out at 9:00 in the morning, and that they had lunch, I can ask them exactly what time they had lunch and find out whether what they ate was good for a post-exercise meal."

Bill has added some features of his own to keep track of his clients. He used to write everything down on paper, but now he has created a spreadsheet on his phone. "I made a template and I chart all the exercises I have my clients do, and I can email that out to them if there is stuff I want them to do on their own in terms of training."

Are Bill's clients good about tracking what they are doing?

"People are held more accountable when they know I am looking," he says. "Granted, I have had clients who no matter what I do, are not going to be held accountable, unless I am there, face to face. But for the most part when they know I am looking, they are pretty accountable."

WHAT TRENDS DOES BILL SEE HAPPENING?

Bill reports that at least half the trainers he knows are now collecting and using some form of electronic data to help them work more effectively with their clients. And he is certain that they are doing it more than they were a few years ago.

He also notes that more phone apps can record exercisers' data wirelessly, without the need for them to carry separate devices that plug into USB ports on fitness equipment. "I think that we will see more apps that will allow users to record that data on their phones without plugging in via a USB. The trend is to make it easier for clients to do it themselves."

"Nutrition and exercise are getting pretty well integrated," Bill states, but adds, "I think the missing piece now is stress management. Figuring out a way for people to collect data on their own stress and manage it would be a major thing. Stress is a huge and often-overlooked part of trying to get yourself into shape."

WHAT IS THE DANGER OF SELF-MONITORING?

If you think briefly about monitoring , you will probably decide that there is no danger in gathering data about your health, medical issues, or exercise. Strictly speaking that is true – you cannot hurt or kill yourself by gathering data. Any danger lies not in the data or the devices that gather it, but in the way you and I think about that data – or ignore it, or misunderstand it, or abuse it, or overlook other key pieces of data about yourself if you fixate on the wrong information.

Take diet pills as an example. You can happily watch your weight drop and actually feel good about that, all while your blood pressure is going through the roof. If you focused on only the weight data, you might think you were fine. In fact, you are not.

We have all tried this quick fix weight loss at least once. I know I did. But then I happened to check my blood pressure one day and it was something like 168 over 105. Wow, way too high. For a man like me, who had a history of minor heart issues and whose father died young of heart disease, that was a really bad reading to discover.

So, you are doubtlessly intuiting my message here, which is that focusing on the data you get from one piece of data-gathering equipment is not going to keep you healthy. It can make you unhealthy, or maybe even kill you. You need a context for data, and an overview of different datums.

A similar thing can happen when people rely too uncritically on sources of medical information they find online. They get a headache, research headaches on WebMD.com, and either convince themselves that they have brain tumors (cyberchondriacs – people who go online and find all kinds of ways to diagnose and fret over ailments that they do not really have) or go in the other direction and conclude that nothing is wrong with them, when in fact there may very well be.

The larger lesson is that gathering and monitoring data becomes safer when that data is monitored by a caregiver/stakeholder, who might also be your primary care physician or a specialist who is helping you get the upper hand over your weight, or your diabetes, or atrial fibrillation, or whatever issue you are trying to monitor and bring under control.

If you are riding herd on a serious condition with the help of a specialist, then chances are good that he or she will already be part of the information chain and will be watching the data that you send. If you are just gathering data on your own without having an expert, interested party review it, you (like I) run a certain danger of making unwise decisions based on inaccurate or incomplete understanding.

Self-monitoring can become something of a hobby in which we use ourselves as experiments. That's a tempting pattern to fall into. But that hobby will be a lot safer when it is done under the supervision of a serious professional caregiver.

Overall, the trends are positive. We're putting more information into the hands of consumers and their caregivers, and that's giving us a more informed, educated and participative consumer.

THE SHRINKING TIME COMMITMENT

I have a friend who developed type 1 Diabetes nearly 10 years ago. At the time, tracking his blood sugar and diet and exercise was, in his words, "kind of like holding down a second job." He took his blood sugar reading every morning, copied it into a little log book that was bundled with his blood glucose meter when he bought it. He recorded his daily exercise and his meals in that little booklet too, using a pencil. Today, his nifty little glucose meter keeps track of all his blood sugar readings and transfers the data to an app that is on both his laptop and his Android phone.

It's all fast, seamless, and immediate. As my friend says, "Just about all the record-keeping is done for me now."

The same is true for most of many other digital devices that have become part of our lives. The heart rate monitor you wear when you go running transfers the data to your app or monitoring platform *in real time*, with no need for you to sit down and write anything down. Given that fact, aren't your objections disappearing? When data-collecting and health-monitoring apps and devices do it all for you without requiring any further effort or time commitment from you, what is causing you to resist? Heck, you are running or going to the gym anyway. You are weighing yourself or cooking healthier meals anyway.

You're putting in the time and effort. Why not let technology do the record-keeping for you? But you and I have to remember that despite all the data, we still need to be proactive in monitoring our conditions and seeking qualified care.

EXPERT OPINION: SEAN DUFFY AND MICHELLE GERALDI OF OMADA HEALTH

Omada Health was founded in 2011 to offer a new way to help people with prediabetes reduce their risk of type 2 diabetes, and to help people who already have type 2 diabetes manage their disease more effectively. Today, 150,000 members are using Omada to lead much more healthy lives.

How do people sign up for the Omada program? They begin by visiting the Omada website, where they answer six questions to find out whether they are at risk for chronic disease. If they are, they qualify to join Omada. A short time later, an Omada digital scale arrives in a box. It looks much like any other digital bathroom scale, but it contains a cellular device that communicates with the Omada care team and feeds data to the participant's private profile.

That begins the process, which moves through two distinct phases of care. During the first phase—called Foundations— members get to know themselves better by tracking and submitting their weight, food, activity and other data. They do so as part of a small group setting and they work with an Omada coach who gives advice on how they are managing their care, and what could be improved. After the first phase, they switch to Focus, where they are coached and supported on hyper-personalized goals based on their progress in Foundations as they make their healthy new habits part of their lives.

The Omada program works. Within the first year of participation, Omada members consistently lose significant amounts of weight. People who are pre-diabetic lose the amount of weight associated with a 30% reduction in the risk of developing type 2 diabetes, a 16% reduction in the risk of stroke, and a 13% reduction in the risk of heart disease. Those are verified statistics, because Omada conducts ongoing clinical studies of members' health.

What was the motivation behind Omada? What difference has the company made in improving the quality of life for its members?

We recently spent an hour chatting with Sean Duffy, Omada's Co-Founder and CEO, and Michelle Geraldi, Communications Specialist. They offered up some first-hand observations about the difference that their company has made in people's lives.

Kevin Pereau: What kind of difference is Omada making in members' lives?

Michele Geraldi: As a former coach, I got to see people avoid developing type 2 diabetes, lose weight, take control of their health, and avoid other diseases. It is inspiring to see so many people benefitting. The word Omada means "group" in Greek. To join Omada is to be part of a group, because our members are not alone in their journey, they are connected to a support community. Many people report that they found us after months and months of searching for a way to take control of their health and finding lots of alternatives that they didn't want to use.

Kevin Pereau: Your website mentions not only diabetes, but other chronic diseases. Are you working only with people who have type 2 diabetes, or do you help with other conditions as well?

Sean Duffy: An enormous body of scientific data shows that pre-diabetes, high blood pressure, elevated cholesterol, type 2 diabetes and excess body weight are all close together, in a matrix. So Omada addresses them all.

The National Diabetes Prevention Program, a famous trial conducted by the Centers for Disease Control, found that when pre-diabetic
people lose weight and make behavioral changes, they have a dramatically lower risk for progressing on to type 2 diabetes. Let me mention
that, as you know, type 1 is a much different disease, genetically based,
that tends to appear when people are younger.

Kevin Pereau: Sean, what led you to start a health care company
like Omada?

Sean Duffy: I was always a tech geek. I grew up in Colorado. My
mom was a nurse, my dad was an engineer. I ended up studying science
undergrad at Columbia University and intended to go to med school.
But when I graduated in 2006, I noticed that at night I was actually
not studying chemistry and molecules, I was reading tech blogs. I got
cold feet about applying straight into medical school. I ended up working at Google for a couple of years. I thought I would do something
in primary care, maybe found a startup. I applied to MD/MBA programs, I got into Harvard, but when I was in that program ended up
doing an internship at IDEO that was a mix of business and medicine.
That was where I ended up meeting my cofounder, Adrian James.

I was determined that it was absolutely possible to help people
with their diseases—prediabetes, type 2, hypertension— and help
them make changes that would reduce the risk. I took a year off from
Harvard and decided that it was possible to help people stop the progression of disease. I saw that there was a way to bring the process to
life and make it both delightful and interesting. So that was the foundation of Omada.

Kevin Pereau: Michelle, what has your story been at Omada?

Michele Geraldi: It's been a long road of seeing a lot of stories, and
working with and watching Omada grow into such an amazing company

Sean Duffy: When Michelle joined the company, there were only
five or six people working here. We now have 200. We have become
the biggest provider in the country and around the world of this kind
of program.

Kevin Pereau: Do you think the social, interactive component of your program has been important to your success?

Sean Duffy: If you are a member, we first give you that digital scale with its cellphone chip. It works right out of the box. You take it out and you step on it. There are no set up procedures to do, no technical headaches. From there, you are matched with a group of 10 other members who have the same health care coach.

The actual program starts on a Sunday - everything kicks off on that day. There is a foundational part of the program that lasts for 16 weeks. Everyone in your group is going through the same shared curricular experience, week by week. And at the end of that, we transition people to a phase that we call a Focus, where you can tailor your program according to your preferences. Perhaps you have a weight loss goal, or you have other goals that can take you a bit further.

But the social component is critical. There are other programs that just give you a calorie counter, or an online coach. But we have learned that if people want to succeed, there are no shortcuts. For most people to make lasting changes, they really must have all the instruments. So our program is fully orchestrated for the user. The first step is simple - you signed up and you started - and then we put you on this amazing machine with different elements that combine in concert.

Kevin Pereau: What challenges have you faced in bringing your company to life?

Sean Duffy: Innovation in the health care sector is amazingly complex. You know, there are studies of financial services that show how complex that industry is, how heavily regulated. But the complexity of health care eclipses that by an order of magnitude.

I sometimes joke that chasing a dollar in the U.S. health care system is like playing a game of Quidditch in Harry Potter, where the players are riding brooms, trying to catch the Golden Snitch, while also trying to score goals. To make progress, you can't go around the system and be disruptive from the side, you have to dive right into it. There are incredible people in the health care system. You have to learn how what you're offering fits in the context of all those stakeholders, and you can't ignore any of them. Omada has to be mindful of what is happening in the diabetes space, the obesity space, government regulators, you name it. We have to learn about all those areas and see how what we are doing fits in and how we impact on all of them.

Michelle Geraldi: Sean has a term that applies to it: "rigorous thoughtfulness."

Kevin Pereau: What business challenges have you encountered as you have built Omada?

Sean Duffy: If you talk to consumer pricing experts, the moment you get beyond a price point of about $100, you are asking people to make a big purchasing decision. Across income levels, you have to find a way to help people who are most in need, and who might not be able to write that check. That has meant ensuring that Omada has been available as a covered benefit on health plans. It has also meant engaging stakeholders like the American Medical Association, all those aspects, to make sure that we will have this covered for our participants.

So if you look at our engineering team and our designers, you see that many of them come from the consumer world, from consumer companies. We have pulled in the best talent from the consumer side and have had to nurture all the different skill sets in one place, and make sure that all those people are really good at getting the best from each other. We really do learn from each other, help each other get smarter.

Kevin Pereau: How rewarding is it to know that you are helping so many people in such a powerful way?

Michelle Geraldi: Well, one thing we have heard often from our members is that they have changed their lives completely. They have prevented or improved their diabetes, done things with their cholesterol, any number of things. At the end of the program, they will tell you that all those things are going much better. That is very common. I have heard it from many, many people. I suppose it's just our mission of trying to enable and inspire people and try to help as many of them as possible.

Kevin Pereau: And that's the answer. Where do you think your company will be in three or five years?

Sean Duffy: In three to five years, I believe we'll be impacting millions of people. Despite all the complexity and challenges of health care, the huge positive side is that you really have a chance to impact the health of people. For people who want the feeling of making a deep contribution to the world, health care is a fabulous place.

THE SMARTPHONE AS A TRANSFORMATIONAL, TRANSITIONAL DEVICE IN HEALTH CARE

If you walk down the street where you live, you will quickly notice that many people seem to be glued to their smartphones. They are using their phones to communicate, navigate, order food and groceries, request rides from one place to another, and more.

A mass migration of health monitoring to those devices is already taking place. That list of different exercise trackers that I offered you a few pages earlier? It could well shrink in size as more and more of the gathering and monitoring functions migrate onto phones. After all, people already own phones, and the cost of most apps is only a few dollars – much less than the cost of separate devices that track exercise, record and analyze what you are eating, and the rest.

Another reason for the migration of health data to smartphones is that smart companies in the Asia-Pacific rim countries have become extremely adept at making and selling low-cost devices that perform the same functions as the expensive devices that are being made in North America and Europe. Many of the presumed North American devices we buy and use are being manufactured in Asia, making it all the easier for companies there to develop competing products.

I recently spent a little time looking over the shoulder of a woman, and talking with her, while she was busy programming a wrist-worn watch that she had bought online for only a few dollars. Did it do everything as capably, or as elegantly, as a Fitbit or an Apple Watch in full flight? No, it didn't. But it was on the way, and you know that many similar products are already being developed and brought to market, just in time to provide some serious competition to the bigger brands we know in the West.

You can pass value judgments on whether that's a good thing or a bad thing from a business perspective. But it is happening, and quickly, and there is no denying it.

Familiarity plays into the equation of how people select and use products, in a way that is like the way people select the cars they drive. If you own and drive a BMW or a Lexus (or an Apple Watch or a Withings watch or a Fitbit), you will love them. If you own and drive a Kia or a Toyota Corolla (and use an inexpensive smart watch you bought online), you are going to love them too. And let's not forget that Kia's and Toyota's are really good cars. The same can be said of a growing number of lower-cost exercise and health monitoring devices. And then there's the fact that quality and features are quickly trickling down from expensive devices to inexpensive ones.

That is always the way the wind blows in the development and market penetration of tech devices. To expect otherwise in the world of health care would be unrealistic.

A CULTURAL AND SOCIAL PHENOMENON

What influenced you to listen to the music you like, to follow your favorite sports team, to live in a certain community, to drive a certain car, to exercise in a certain way . . . or to make all the thousands of other choices and decisions that have shaped your life?

Your first response to that question might be, "I choose what I do, I make up my own mind." But I'm willing to bet that even though all of us are independent thinkers (or would like to believe that we are), we have been heavily influenced to make our choices. The decisions we make regarding what we choose to do and buy are limited, at the most basic level, by what is available. You bought and love your new curved-screen Samsung television, but you could only choose it because it was available.

I know it might sound somewhat blasphemous, but a "birds of a feather" force also affects our choices. I go to a gym and have gotten to know other people who go there. We talk about new discoveries we have made about how we exercise, monitor our fitness, and lots more. There are social influence inputs that exert a transformational force on what I do and how I do it.

This is the "Cheers" effect that I mentioned earlier, named after the famous television show with a theme song telling us we want to go to a place, "Where everybody knows your name . . . and they're really glad you came." Online communities are all about content, context, belonging, and social influencers.

Our eagerness to apply digital health and wellness data is being driven by social influences. The digital health drivers know that ultimately, social forces will move acceptance of devices from the periphery to the center of people's lives. Some think it already has.

Lots of technology is available for you to use. Your physician gives you a monitoring device and you take it home. You visit an exercise club that has equipment that remembers and tracks your previous workouts. You download an app on your smartphone that lets you track your meals and calories. But what gets you to decide to actually use one of those pieces of technology and make it part of your life and your routines? Nine times out of 10, it is people that do.

MERGERS, ACQUISITIONS AND MEGA TRENDS

As a consultant who frequently focuses on building health care platforms for aggregating, monitoring and analyzing data, I have noticed a few other important trends.

A big one is that traditional business categories are blurring. Restaurants have apps – that is already old history. Pharmacies are giving shots for the flu, shingles, pneumonia, and other diseases. Dieting and nutrition companies that offer weight loss assistance are acquiring or merging with exercise chains. Insurance companies are working with app developers. Retail pharmacies are jumping into the insurance business. Insurers are integrating products, programs and services that allow them to better span the health care value chain.

Only a few years ago, who would have thought that such changes would be possible?

Dr. Eric Topol, a physician who has written books like *The Patient Will See You Now*, believes that we are seeing a transformation in health that will change the world as much as Gutenberg's printing press did back in the 15th century. That put the printed word into the hands of common people for the first time ever. This is much like what smartphones are delivering to a health care consumer today. Before long, new technologies will put Watson Supercomputer-level analytical capabilities into the hands of everyday medical consumers like you and me.

The more we collect, the more we can analyze and predict. The faster we can intervene, the better the expected outcome. Blockchains are breaking data down to transactional bits that enable secure sharing, analyzing and prescribing. Welcome to connected health!

EXPERT OPINION: JOHN L. BROOKS III, CEO, HEALTHCARE CAPITAL LLC

Diabetes is an epidemic, with tens of thousands of new patients developing the disease every year.

Perhaps more than any other condition, diabetes requires people to gather, monitor and understand statistics about their own health. People with type 1 and type 2 diabetes monitor their blood glucose levels, the results of their A1C tests, and the nutritional components of the food they eat. Many people who are newly diagnosed with the disease feel like they have been thrown into a new world of numbers.

But what if new solutions came along which, by gathering and communicating data, made it easier for people with diabetics to monitor and manage their health? What if instead of concentrating on numbers, people were freer to simply exercise, eat well, and focus on healthy behaviors?

John Brooks is the CEO of Healthcare Capital, LLC, a consulting company that advises companies that are developing advanced, disruptive health care solutions, especially for diabetes.

Over the years, John has co-founded seven life sciences companies, including Insulet, developers of the Omnipod® insulin-delivery device. He is also co-founder of Prism Venture Partners, a $1.25 billion venture capital firm.. John has previously served as Chair and then President and CEO of the Joslin Diabetes Center, a Boston-based diabetes research, clinical care, and education organization. Earlier in his career, John was the Emerging Business Group General Manager at Pfizer/Valleylab, and President/General Manager at Pfizer/Strato.

Today, John is Chairman of the Board of Cellnovo Group SA, a French company that is developing pioneering systems for mobile diabetes management. John is also the Chair of Thermalin, a Cleveland-based company working on next generation insulin analogues and Chair of the College Diabetes Network, a non-profit organization that provides peer support to young adults with type 1 diabetes on college campuses across the US. He also serves on the board of Hygieia Inc., an Ann Arbor-based company that is developing new digital insulin titration systems for people with type 1 diabetes.

John's achievements, impressive though they are, don't tell the story of how he came to care so deeply about people with diabetes, or why he has devoted his life to combatting the disease. I am honored to share these excerpts from a conversation I had with him recently.

Kevin Pereau: How did you come to do so much to help people with diabetes?

John Brooks: My involvement began 25 years ago, when our three-year-old son was diagnosed with type 1 diabetes. At that point, I made it my life mission to understand as much as I could about diabetes, and to try to make an impact on the lives of people who were living with it. I have spent the last quarter century overseeing diabetes companies, investing in diabetes companies, advising diabetes companies, and founding them. I had the opportunity to run the Joslin Diabetes Center, and now I am on the boards of several companies in the diabetes space.

Kevin Pereau: There is so much happening with diabetes care. Where do you think the greatest potential lies to improve care?

John Brooks: I like to get involved in companies that have real transformative potential to impact the delivery of patient-centered diabetes care. The key issue is that we must confront the lack of effectiveness and efficiencies in the current health system and we need to make investment and improvements that will help us get ahead of the global diabetes pandemic that we are facing.

I focus on companies that have real disruptive capabilities and that are bringing new solutions to advance diabetes care.

Kevin Pereau: What important trends have you seen come along in the last decade?

John Brooks: I would be remiss if I didn't mention Insulet, a company that a partner and I started. We were co-developers of the disposable OmniPod® device, which delivers insulin as needed, when needed. I am now building on that theme with my efforts at Cellnovo, where we can use Bluetooth and sensor inputs to support real time, "closed loop" insulin and diabetes care optimization. The data analytics/machine learning aspects of these systems will provide predictive and adaptive insights that will enable people with diabetes to live healthy and productive lives without the worry and hassles of today.

Now, we have wearables, sensors, novel delivery systems, "Internet of Things" devices, and all kinds of apps and solutions that can be used to help people manage chronic conditions like diabetes, and other co-morbidities such as heart disease. Yet to make progress we need to move beyond thinking that we can just throw technology at health care problems. In the pre-digital health era and even today, the doctor had only a short time to try to figure out what was going on with a patient. Depending on the kind of specialist that physician was, he or she would map out a care plan, write a prescription, send the patient home, and then schedule another appointment 90 to 120 days out. In the interim, there wasn't a lot of involvement or interaction between the physician and the patient. Unfortunately, the next inter-action could occur when a patient showed up in an emergency room.

That is now starting to change. Physicians are increasingly able to gain insights that have been derived from the ongoing data coming from their patients. There is more personalization. They can monitor patient activities, medication adherence, and then provide meaningful and timely feedback. If a patient is not taking their meds, or not taking the right amount of insulin, or not exercising, the care team and the family will know in real time and can provide feedback via a text mes-sage, an email, or a call from a coach. The cloud-based platform could also send an alert that goes to the patient's family members.

What I like about the last 10-15 years is that individuals are being able to get recommendations and feedback, making it easier for them to engage more fully in managing their condition, but in a way that suits them. They understand they have a chronic condition and they want to know how to improve their health without having their con-dition overwhelm them. This is an entirely new way of getting better outcomes from connected care.

You know, when the health care industry started out gathering data, they thought that if they just pushed it out to doctors, clini-cians and their teams, they would figure out how to use it to improve patient care. Now we know that will never work, and it will never result in better care. We are now able to gather data and turn it into what I call actionable, intelligible information that can help inform the patient and the care team. Thanks to artificial intelligence, the use of databases and algorithms, and other advances, care teams and health coaches are being alerted if they need to intervene in a patient's care. Plus, there is the important psychosocial aspect, whereby the patient's friends, family members and social circles get involved in helping the patient to do what he or she should.

We are beginning to understand how we're going to achieve better clinical outcomes and by measuring them, these improvements will lead to cost savings, cost avoidance, and better overall care.

Kevin Pereau: Will the gathering of patient data lead to better outcomes not just for individual consumers, but also for groups of individuals who are dealing with diabetes and other diseases?

John Brooks: I was just reading about a pilot study that United Healthcare and Dexcom are about to do. About 10,000 people with type 1 diabetes will be given continuous glucose monitoring devices and their diabetes improvement results will be studied. It should be a real eye-opener. It will collect data about what happens to glucose levels when people in the study are sleeping, and what the effects of stress and exercise are. Patients will get a sense of, "this is me, and this is my personal disease," and not something a patient never felt was relevant to them, based on general data or statistics.

We're just starting to appreciate how we can use AI-enhanced technology to be far more proactive about helping to alleviate the challenges of people dealing with a chronic disease. We might even be able to start thinking about getting ahead of the curve by taking a more proactive role in helping pre-diabetic individuals; instead of making vague recommendations like "lose weight" or "eat less." We will be able to identify people who have a genetic predisposition for developing diabetes and give them a wellness road map and recommend specific programs to encourage them to do the specific things that will help to prevent diabetes.

We're just starting to understand and appreciate how to involve consumers, employees, patients, care providers, and communities to prevent and alleviate chronic diseases. We are starting to genetically flag if patients have a disposition to develop diabetes. The day will come when we will know which restaurants they're going into, what they are eating, how much they are exercising, and have an opportunity to encourage them to do the things that will keep them well.

Kevin Pereau: That kind of proactive, preventative care is a far cry from waiting three months between doctor visits, and then discovering that something has gone wrong.

John Brooks: Yes, in the past we were applying an acute care model to a chronic disease. And as you know, chronic diseases are not best treated by a number of episodic meetings. With actionable, intelligible information, we are more able to help people deal with chronic conditions, 365 days of the year, 24/7. Yet we have a lot further to go. If someone is only seeing a care provider two or three times a year, we have the opportunity to do better and provide much better care. Remote monitoring, virtual visits, and smart devices and apps will make a real difference.

Kevin Pereau: I know that technology is changing fast, and everything we discuss today will probably change in a few years. Yet nonetheless, are there trends you would like to say are important?

John Brooks: As we have highlighted, technology is changing dramatically. You have Apple, Samsung and other consumer electronic or consumer-facing companies coming more into the diabetes space. We are seeing new apps and ways the Apple or Samsung Watch and mobile devices can be used. Google is developing miniature sensors that can be used to gather patient data. It seems that every major company is looking at the consumerization of health care, and Amazon has been suggesting that they want a greater role in health care. And clearly, more consumers are monitoring their health, exercise and diets and more on their smartphones. And storing and sharing data on the cloud is another part of it.

There is an opportunity to transform all this data into intelligence that can be used to help people be healthy. Digital health care is exploding. It seems like there isn't a day that goes by when something new isn't being announced. And more and more pharmaceutical, retail, payer, device, and provider organizations are recognizing that they need to be part of this revolution. They can no longer just think about where their customers are going to buy their next vial of insulin or strip. And the FDA is slowly but surely understanding that they must accommodate all these changes.

When you look at where my son started out with managing his type 1 diabetes, he was taking multiple injections of insulin daily. He was using meters that took about 30 seconds to do a blood glucose test. That was pretty challenging. We've come a long way, but we have a lot further to go. It requires thinking out of the box about people with diabetes and finding ways to help them during the course of their disease. And not just thinking that with a little bit of advice delivered once or twice a year, that patient is going to do a good job of managing a condition on his or her own. That was an approach that we can now improve with the appropriate deployment of tools, apps, solutions and devices, and the cloud, to help people over the rough spots and help them keep moving in the right direction.

Lilly just announced a strong desire to be in the digital, connected health space. Other drug companies are starting to realize that they need to provide complete one-stop solutions. They can't just be thinking about selling a product. They are seeing that they need to help patients arrive at better care through cost effective solutions.

I believe that the providers' economics, and how caregivers are compensated, should be focused on outcomes and on how well people are doing. We need to reward doctors for using digital tools. There are opportunities to engage patients at home, through telemedicine if you will, so we can allow people to be much more engaged in their health care but in a way that works for them.

Kevin Pereau: What do you think the larger impact will be of the merger between Aetna and CVS?

John Brooks: It could be transformative. When I had discussions with CVS many years ago, their plan was to have walk-in clinics that dealt with fairly minor issues. If somebody had a sore throat, had a rash, needed a flu shot, they could talk to a pharmacist or nurse practitioners at a CVS location. But I think the company now sees there is an opportunity to help people with chronic conditions including diabetes, congestive heart failure, and other long-term conditions. CVS may well be adding more health care professionals to their staffs.

If you think about diabetes, there are no "procedures" that need to be done, *per se*. Treatment is mainly about helping people understand the complexities of the disease and how various medicines, devices, lifestyle changes will help impact their condition. And we are discovering more and more that diabetes is not just about glucose. Thanks to the gathering of more data, caregivers will be thinking about sleep, stress, hormones, co-morbidities, and all the other factors that affect glucose levels.

It might be speculative but look at the transformation that Amazon has made in the retail space. A lot of people are thinking that similar changes are about to occur in health care, which has generally been behind the times. It was a major step forward to move from paper health records to computerized electronic health records (EHRs). We are now starting to ask bigger questions of the EHRs. In the past, the cardiologist would manage a patient's heart disease and the endocrinologist would manage their diabetes. And now, thanks to sharing of patient data, there are a growing number of instances in which these specialists are talking to each other.

I love what I am doing. It is a great time to be in health care. I am encouraged by the fact that we are starting to embark on new ways to use analytics to drive real change, instead of just kicking the can down the road the way we have been in the past.

CHAPTER SEVEN

BRIGHT MINDS, BIG IDEAS, NEW VENTURES AND SUCCESS

Things are changing quickly in the world of digital health. As soon as you think you have discovered the most promising new company, idea or technology, something newer comes along.

So, where can you go to learn, stay ahead of the trends and discover the most important new companies?

I can only observe that there is no one blog, publication or other source that you can visit daily to keep up-to-date. But there is one thing you can do:

Keep an eye on what is coming up at digital health expos and conventions.

In this chapter, I would like to introduce the viewpoints of Constance Sjoquist of HLTH.com and Matthew Holt, founder of Health 2.0. Both their companies organize annual digital health trade show/conferences. Their conferences are, in fact, astonishing events where innovative new health care companies come to display and talk about their cutting-edge technologies.

EXPERT OPINION: CONSTANCE SJOQUIST OF HLTH.COM

HLTH.com, a company founded by Jonathan Weiner, is reshaping the industry narrative on what it means to drive reductions in health care costs, continually innovate and increase quality in health care.

Their conferences are all about uniting people from all phases of health care: insurance companies, hospitals, benefit professionals, employers, venture capital investors, start-ups and more. Their typical list of presenters often includes executives from 23andMe, Accenture, Aetna, Amazon, Aspire Health, Boston Children's Hospital, the

Cleveland Clinic, Color, Comcast, CVS Health, Fidelity Investments, Fitbit, Frost & Sullivan, Google, Kaiser Permanente, Garmin, GE, Humana, LinkedIn, Microsoft, Omada Health, Optum, Uber, United Health, Walgreen Co., Walmart . . . the list goes on and on. They pull together thought leaders and host great events.

I spent some time chatting with Constance Sjoquist, Chief Content Officer, HLTH, and here is what she told us.

Kevin Pereau: How exciting are HLTH events, and why?

Constance Sjoquist: Very exciting! From a historical perspective, I think we are at a point where many different companies each have their own Swiss Army knife, filled with tools. But people are looking at other companies and saying, "Even though what you have is really cool, your can opener doesn't open my bottle."

Or they are looking at other companies and saying something like, "I have too much invested in manufacturing, and there is no way I can align with you, you are just too different." We are thinking that HLTH will be a place where people will rearrange how they think about their technology, talents, investments, and discover one another and use what they have to their advantage.

Kevin Pereau: The Swiss Army knife analogy is very fitting . . . how does data fit into it?

Constance Sjoquist: I think that in the past, people have been talking about data as though it is some kind of flowchart, or maybe a warehouse storing lots of information. But data is essentially nothing in and of itself, right? It's just ones and zeros and stuff. It's like the stuff I put in my dresser drawer and I don't know why I'm keeping it.

The data that many organizations have gathered is like that. It's hard to throw it out or to clean it out, or to understand how to share it. You're thinking, "No, I don't want to give it away because I may need it someday . . . If I give more than I have, I'm left with nothing."

So there is competitive worry around data sharing, privacy, and security. I think that is pretty much the old thinking and the old conversation that was taking place within organizations.

Kevin Pereau: And what is the new conversation?

Constance Sjoquist: There are many taking place I think, but I think the big question that is being asked is, "What do I do with the rest of my time when I am not in the compensation bucket?" I think that people are thinking about the whole orchestration of a larger eco-system around health.

Kevin Pereau: What are some of the tech trends that you have discovered in the companies you are working with as you plan?

Constance Sjoquist: I think that there will be conversations at HLTH about developments like 3D-printed body parts that have intelligence built into them, and about the increasing use of genomics in predictive diagnostics, based on the growing realization that people respond in different ways to different medications, and that they are more or less likely to develop certain diseases and conditions, based on their genes. I also think there will be many conversations about retail, and who is getting into the retail health care space.

I often get calls from people who come from outside of health care who say, "We have a payment solution" or, "We have this great soft-ware" or even, "We have a bunch of people in our company who are entrepreneurial, but they don't know how to succeed in health care, how can we figure out what to do?"

The Geek Squad concept works with computers and home elec-tronics. You buy a piece of equipment, you have a problem, and the Geeks come. But in health care, the Geek Squad just doesn't work. We are talking about a whole interconnected ecosystem that is made up of so many different entities and kinds of companies.

It is time to share. In my view, sharing is the new "sprinkle dust" that is getting mixed into the industry.

Let's meet Matthew Holt, another observer who is uniquely posi-tioned to have an informed opinion on recent trends in health care.

EXPERT OPINION: MATTHEW HOLT, FOUNDER, HEALTH 2.0

Matthew Holt founded Health 2.0 with Indu Subaiya in 2006. Today, the company occupies center stage in the world of digital health. Its annual Health 2.0 conference in San Francisco brings together representatives from every important company on the digital health care landscape. The 2017 conference, for example, was attended by more than 150 companies that gave live demos of new technology. There were four keynote speeches by industry leaders and three CIO interviews. And more than 10 new companies announced their startups at the conference.

Health 2.0 now runs conferences around the world. Matthew also started and contributes to The Healthcare Blog, one of the most widely read blogs about wellness and health care technology.

Few people have Matthew's broad perspective on what is happening in the world of digital health care. I thank him for adding his deep insights to my book.

Here are some outtakes from our recent conversation.

Kevin Pereau: Where did you come up with the term Health 2.0?

Matthew Holt: Ten or 12 years ago, people started using the term Web 2.0. And people piled on and whatever was new, they started calling it something 2.0. And that's where our Health 2.0 name came from. We thought we would have to change the name of our company and our conference to reflect changes in the health care world, but we have kept it the same. We have kind of adapted the definition of Health 2.0 to fit what is going on.

The term digital health has also been around for a while. It got picked up in 2010, 2011, 2012. People started referring to the changing world of health care as "digital health." My concern is that the term is wrong, because "digital" implies something that's on a computer. What is really going on in health care now is, especially in the U.S., that we are basically building out enterprise-level electronic medical records.

We have a way to go. Look at what just happened after Hurricane Harvey hit Houston. Think back to Hurricane Katrina in 2005, when hospitals learned that they had to keep and protect their medical records. Yet when Harvey struck this year, the event was of such magnitude that people ended up in hospitals that didn't have their records. That is another way of saying that their records were enterprise-level, not consumer-level.

When you think about that, that's kind of where email was 20 years ago, right? We had email servers in our offices and that was where files were stored. Now email is all in the cloud. Medical records are heading to the cloud too, but that same level of change hasn't happened yet in health care.

I think that the term "digital health" summarizes all that change. But I like to call what's happening today SMAC health. I mostly use the term to tease people. It stands for Social Sensors, Mobile, Analytics and Cloud. To me, that's the right way to think about what's happening at the moment, because that's the new stuff. It's being added to by all the newest technologies like AI, virtual reality, meta reality, block chain. There's a bunch of new stuff coming down the pike, which is experimental.

We don't have mainstream driverless cars yet, and we don't have mainstream AI yet, but we do have mainstream cloud-based email! But I think that if you start trying to use the terms Health 1.0, 2.0 and 3.0 to try to figure out how they are different or when they began, you are only going to get confused.

Kevin Pereau: Yet there have been some seismic changes along the way, correct?

Matthew Holt: Well, let me riff on this question. An important stage occurred when people began accessing and sharing information and data with other people. That coincided with the rise of WebMD and patient communities. But all that stuff had been festering in the background for years, all the way back to Dr. Spock telling people how to look after their kids. And to the Boston Women's' Health Book

Collective, writing the book *Our Bodies, Ourselves*. Then you look at the AIDS movement, the collective sharing of information about cancer, right?

After the Internet exploded, by the late 90s, you had thousands of forums and user groups for people discussing health issues. There are thousands of forums today on Facebook alone. By the mid-2000's, there was shareable data. Patients Like Me, which has been around since 2005 or 2006, represented a major change because it was not only people talking to each other, but beginning to share their data with each other. It's an amazing service.

And today, there are lots of people who are starting to track themselves. Not necessarily people who are dealing with medical issues, but healthy people too. They're starting to use analytics to try to figure out what's going on with their health. That trend is certainly growing, and it will grow even more.

Kevin Pereau: What new developments are coming up?

Matthew Holt: Well, you have a number of things. There will be virtual meetings with doctors . . . synchronous and asynchronous tools to connect patients with doctors, to replace face-to-face visits. So that's a big deal.

Another big thing is the prevalence of sensors in the home, and on patients themselves. You're already starting to see companies with body patches and socks, and God knows what, that are tracking everyday activities. There are the fitness freaks who are wearing devices like shirts that track their movement. And there are people with diabetes who are wearing socks that measure the temperature of their feet and alert them if they have an ulcer coming.

Then you have companies, including Nokia and Philips, that build thermometers and weight scales and blood pressure cuffs and devices that patients can breathe into to measure lung function, and devices that can alert patients if they have heart disease and monitor them if they do.

Kevin Pereau: I have a friend who has atrial fibrillation and is wearing a heart monitor right now. He'll wear it for two weeks.

Matthew Holt: You're talking about a CardioNet monitor? CardioNet was one of the earlier companies in cardiac monitoring. Often, patients get sent home from the hospital with their devices. Those devices collect tons of data and communicate it constantly to a central CardioNet location. That kind of service is only going to expand.

Between weight, blood pressure, diabetes, there will be a lot going on. And at-home diagnostic tests are increasing in number. You can pee or poop and send a sample off through the mail. You can put a bit of blood on a stick or on a sensor and send the data remotely. At Health 2.0, we even had an entrepreneur who had a service where a user could put a sample of his semen on a slide, take a picture of it with his phone and get a report about the mobility and motility of his sperm. The service then sends him advice on what he can do to increase his sperm count!

So all that stuff is coming. Generally, technology is adopted first by sick people who have conditions to test or monitor, and then the tech spreads to other users from there. Self-help types like serious athletes love this stuff, because they can measure themselves continuously, as opposed to taking point-in-time measurements only at the doctor. Elite athletes really want this technology, and they often don't care how much it costs.

People with diabetes have been pioneers. In the past, there were many ways for them to collect their data, but more difficult for them to communicate it to health care providers. But thanks to new glucose meters and apps, all that data is now getting aligned between patients with diabetes and their doctors. And it is a very big thing for moms and dads who have children with type 1 diabetes. They want to monitor their kids' blood sugar levels. And while they are at it, they can access other information too, like their kids' immunization records.

And then of course, you have people who have long-term chronic illnesses like cancer. There is a lot of activity around them.

Frankly, for the 60% of the population who don't get sick, all this technology probably doesn't matter that much, and they don't spend money on it. But for people like me? Maybe because I'm in the business, I step on a scale every morning and my data goes into the cloud and I say, "I am five pounds overweight again? What happened, Thanksgiving?" But my involvement is less than it would be if I had a chronic illness.

Widespread adoption of what you, Kevin, like to call "digital health assets" is just getting out there. But compared with consumer acceptance, it's had a kind of limited provider acceptance. But it's getting more and more there. More trials and providers are starting to push it.

I believe we are at the early stages, when the S curve starts curving upwards. We're at that early point where it just starts to curve up. Frankly, it's not going to curve up as quickly as the Internet did, or cellphone usage. It's going to curve up on a languishing, lazy S curve, but it's going to get there.

Kevin Pereau: What new developments do you expect to see at your Health 2.0 conference this year?

Matthew Holt: I think there is a growing interest in genomics, as people get the tools to look at their own microbiome data and can understand the impact on the drugs they take.

In Santa Clara, we're going to have demonstrations and conversations from some of the big players. One trend is that companies are putting more medical records into the cloud. And more companies are beginning to build applications that will be built on top of those platforms, in much the same way that developers built applications on top of the Android and iOS platforms. That's just coming to health care.

And there's a very important underlying standard called FHIR [Fast Healthcare Interoperability Resources] which we'll be featuring, both in a technical developer way, and in a general discussion.

And then you have this new experimentation with block chain, which is another way of allowing people to access data and to make small automated commercial transactions and contracts, so people can start paying each other and moving data around while verifying the identity of patients and doctors. The motivation for care providers is to lower transaction costs, as more of the transactions that were done manually will begin to be done automatically. Block chain is one opportunity to develop an underlying technology for doing those transactions.

I'm running a panel at the conference on the future of care providers. I'm focusing particularly on the technologies behind diabetes and cancer care.

Kevin Pereau: How do you think all this will impact the lives of patients and consumers?

Matthew Holt: One thing is that big companies like Microsoft and others, as well as several small startups, are developing ways to predict who will become ill, and to be proactive around that.

Companies like MySugr and Livongo and GlucoMe are developing new tools to manage diabetes. They have sensor-based glucometers that communicate with coaches and with the Web and which tell people whether they have an issue to deal with and what they should do if their numbers are out of whack.

Another interesting company is Virta Health, which has a virtual clinic to reverse diabetes. They offer tools—including an app and remote monitoring by physicians—that can help patients get their numbers down to a pre-diabetes state. They are working with what is called a keto diet – which centers on protein and healthy fats.

Similar projects are underway around cancer care. You have a lot of companies, Flatiron Health is one, that are developing new ways to put together what might be called "full stack" cancer care integration.

More than 150 companies will be coming to the conference. We have hospitals coming. Tons of stuff. It will change lives, and it is good fun.

CHAPTER EIGHT

DOING WELL BY DOING GOOD - REBUILDING THE CONSUMER/PROVIDER BOND IN THE NEW WORLD OF DIGITAL HEALTH

One of the more gratifying aspects of writing *The Digital Health Revolution* has been getting to spend time with the thought leaders who are driving the disruption that is simplifying health care. Health care is becoming more of a consumer experience thanks, in part, to their efforts.

While it's fair to say that one of the early myths about digital health was that consumers won't engage or share, it is equally fair to say that a continuing myth is that nobody is doing anything about it. Plenty of people in health care recognize the magnitude of our problem and have dedicated their lives to making a difference. It is uplifting. It is fascinating to see the industry show appetite for better connecting with the wave of innovation that is taking place. In the early days of digital health, the push back for our health care problems always seemed to be lay blame on consumers. Now, we are experiencing tailwinds and a desire to connect with them. We are addicted to our smartphones and we respond to our health care notifications the same as we do for our Facebook, LinkedIn and Twitter notifications—with gusto! We are all in. Consumers are speaking loud and clear and the health care industry is listening. Health care is becoming a more consumer experience, and that means that insurers and providers have to get to know us better and use that data to help us achieve our health care goals.

Not only are consumers speaking up, they are asserting control, saying what they want and how they will consume it. It wasn't that long ago that patients with debilitating illnesses or chronically monitored conditions didn't have much help outside of their doctor visits. Today, companies like Omada are talking about being able to reverse type 2 diabetes. Well, if a company like Omada is typically sold through your insurer to your employer, how would the average person even know about them? Amplifying the voices of those driving the disruption is what motivated me to write this book. Their entire missions are based on delivering better results and outcomes and most people have still never heard of them (yet).

We are now seeing employees encourage their company's benefits department to explore ways of providing tools focused on helping us stay healthy, manage chronic conditions or simply navigate the complexities of our health care system. We have seen examples of who is emerging as market leaders. Rally Health's mission includes making shopping for your annual health care plan and finding the best doctor as easy as shopping online at Amazon. What's more, their corporate wellness offerings help tie everything back to your benefits plan.

Who wouldn't want to see choices like that during benefits enrollment? If your company isn't evaluating solutions that can help you simplify your health care shopping needs, it should be.

The same can be said of innovators like VIM, Omada and others we checked in with for this story. If your company can save money while helping you stay healthy, don't we all win? The advances in digital health over the last eight years have been simply astounding. Continue advocating for yourself. Everyone wants to connect with the consumer—speak up and show them how.

It is gratifying to see innovators, insurers and providers all rowing in the same direction, but I still worry about whether we are reaching those most in need. The first wave of digital health innovation focused on helping those with chronic conditions better manage their conditions. There haven't been a lot of companies focused on addressing social determinants of health, those who can't afford or simply can't get to and from their care provider. Problems like addiction, depression and mental health have yet to see the traction with digital health assets that better connect them to the resources, content and social networks they need for their recovery. It is ironic because this demographic probably most inspired recent changes to how providers are reimbursed for Medicare, Medicaid and Children's Health.

Who is advocating on behalf of the poor, the geographically challenged and those most in need - but most difficult to serve? How do we get investors, innovators, insurers and providers all in for this market segment like they are for others?

There are three key fundamental performance indicators to measure in health care. They are access to care, affordability of care and quality of care. How do we ensure we don't leave anyone out?

This next section features some of the most inspirational people in health care. While others have mastered the fine art of broadcasting health care's problems, meet the people who are not shying away from solving some of the most daunting challenges we have in health care. Many will see their mission driven initiatives as a moral imperative but it is also important to note they are also driving down overall costs by helping to keep an important and often neglected demographic healthy.

EXPERT OPINION: ANDY SLAVITT, GENERAL PARTNER, TOWN HALL VENTURES

During his career, Andy Slavitt has led many initiatives that have shaped our national health care system and improved the lives of millions of Americans.

In the 1990s, Andy was the founder and CEO of HealthAllies, a digital health company that served the un- and under-insured. He then he spent a decade at Optum, which he conceptualized and led from before its launch until it became one of America's most important insurance providers with an annual revenue that exceeds $80 billion. Andy next served as the Acting Administrator of the Center for Medicare and Medicaid Services (CMS) under President Obama.

As we write this book, Andy is General Partner at Town Hall Ventures, a venture capital group launched in 2018 with the high mission of building companies that provide needed health care to underserved populations. Town Hall Ventures's portfolio companies include, Bright Health, Cityblock Health, Somatus, Strive Health, Landmark, and Welbehealth.

In light of his long and varied leadership roles, it is hardly surprising that Andy has become a highly recognizable authority on health care, with a column in *USA Today* and regular appearances on CNN, MSNBC and Fox. He was recently recognized by *Modern Health Care* as one of the 10 most influential people in health care.

We are very excited that Andy took the time to share his opinions about where health care has been, and where he believes it should be going next.

Kevin Pereau: What led you to start Town Hall Ventures?

Andy Slavitt: We are now, as a society, in something that is called digital health, but back in the 1990s, it was called dot.coms. I was in the private sector then, running a large health organization called Optum, which today is an $80 billion business.

Coming out of experiences like that, I felt that I perhaps had a lens that allowed me to focus on the most pressing problems to address in health care, those that needed the biggest focus. When it came to innovation, my point of view has been that many entities in health care have invested a lot of money in places that do not really move the needle on improving care for the people that need it most and spend a lot of our resources in places where we really don't need to.

Kevin Pereau: Where should we be investing?

Andy Slavitt: Where do we really need to invest and innovate? Let me point out that we have 120 million people in America who spend about $1.3 trillion on health care every year. Generally, they have very ordinary needs, with relatively the same diagnoses that you and I have. We have health care issues that are being relatively well served. Instead of investing in companies that address the needs of underserved people, many venture capitalists and other investors ignore them. I see some companies where there are groups of 35-year-old men who are deciding that it would be good to invest in technologies that appeal to people just like themselves, or to 70-year-old men who go running every day while wearing two Fitbits. And those people are not the root of the health care problem in America, or the biggest or most important market.

We have to think about the woman who lives two bus rides away from her dialysis treatment center, or she can't get there. She misses the bus, or the scheduled hours of her treatments don't work around her job, so she goes into renal failure and spends a lot of time in the hospital and probably dies. Now, many of the investors I see are perfectly empathetic to a woman like that, but they aren't thinking about her. They look at the problems through their own lens, and are thinking about the wrong needs of the health care system.

Another issue is that when we are trying to improve the overall system, we often focus on assessing the needs of companies that provide coverage for their employees. In other words, when we set out to fix or invest in the overall care system, some of us think only about the needs of employer-based coverage.

That could explain why to date, many investments haven't generated significant ROI for investors. We have been focusing on the wrong part of the problem.

Our view when launching Town Hall Ventures is, let's innovate in creating opportunities to care for people in ways that consider their needs and the relatively extraordinary lives they lead, and produce real and needed change. A minority of the people in the U.S. are great at being self-advocates. Most Americans do not eat in a healthy way or exercise the way they should. They can't afford their prescriptions. They have so many barriers to overcome. They can't find quality health care in their neighborhoods. They're single parents. They can't find or afford good housing. When they need behavioral counseling, there is no way to find that where they live. So a number of challenges exist because there is no easy access to resources.

That is where health care costs explode, and that is where we are looking to invest. We want to invest in and build 15 great companies that are involved in improving the way care is delivered for large pieces of the population. Underserved populations. That could mean frail, elderly, hospitalized, disabled. It could mean that they need intervention in specific areas like maternity care. This is where Town Hall Ventures sees real opportunity to invest. We see some great companies evolving in this space, many of which are community-based, and we are looking to help build those companies. That is why we started Town Hall.

Kevin Pereau: How do you get others to care about what you are doing and share that viewpoint, perhaps to see health care as a moral imperative and a universal right?

Andy Slavitt: You have to make the business case for it to be doable and repeatable.

If you look at Medicare, over the last 10 years, we have built in more incentives to make it work. Think about since Medicare Advantage, all the innovative companies that have been launched to serve the population, and how primary care services are redesigning themselves. No one cared about the Medicare population until MMA[1], right?

Something happened that disrupted, and made it change. You have CAHP Scores[2] and all these things, and so that happened in Medicare. The same thing is happening in Medicaid. Seventy-five percent of Medicaid is now managed care.

You talk to hospital CFOs and you say to them that 30% to 40% of the patients they serve are on Medicaid. It is the lowest paid book of business. But you can then say to them, "How much do you lose on your average Medicaid admission?" Most of them will tell you something like $5,000. Every time they admit somebody into their system who is on Medicaid, they lose $5,000.

---

1 The Medicare Modernization Act, passed in 2003 - ED.
2 Consumer Assessment of Healthcare Providers and Systems, a rating scale introduced by the U.S. Department of Health and Human Services, in 1995 - ED

One of the best ways to make money is to lose less money. So if you can prevent those individuals on Medicaid from getting admitted for something that they could find treatment for in their community instead of in the ER, hospitals can lose less money. So even if you are building these models for better care, incentives are already built in. There are already incentives built in to help make that change happen. So people who are caring for this population understand its suboptimal opportunity. It is being addressed right now.

So the question is, how do you attract investment in order to address that kind of equation? People will invest if they see ideas and solutions. And then another question is, how do you attract entrepreneurs like Cityblock Health, Omada, and Landmark, which have all launched in the last few years?

What traits do the people at those companies have in common? Well, many are entrepreneurs who are probably less than 40 years old, maybe 35, 37, 38. Incredibly mission driven and smart, tech-savvy, and what they deliver is enabled by technology. Yet the smartest of them don't expect people's lives to change just because they developed an app. They are actually using technology to deliver better care.

They're funded by really strong venture firms. Landmark is probably the best example, because they're furthest along. They're aggressively growing and going into people's homes and changing the way care is delivered. Plus, they are getting superior outcomes.

And then once that that happens, lemmings appear. There are probably 10 versions of Landmark out there right now. We are receiving tons of business plans from people who want to do what Landmark does, and to be that.

And so overall, we're not having as much difficulty pushing uphill as it once was. People are increasingly focused on getting at these issues. And I think that the fact that investors have been supporting a lot of entrepreneurs who are focused on the right things is what all the difference has been.

Kevin Pereau: How do you differentiate Town Hall Ventures from other firms that invest in health care? Is it the partners and alliances that you have? If I'm sending a pitch sheet to Town Hall vs. other venture investors, what will I see that is different about the way you consider my proposal? Why do innovators want to work with you?

Andy Slavitt: So, we have a different approach. Number one, we are stage agnostic. We will look at anything, from an idea on a napkin to a fully developed company that's in a growth stage. But it's got to be mission-specific. Because of Oxeon Partners (Town Hall General Partner Trevor Price is CEO and Founder of Oxeon), we can build great management teams for lots of entrepreneurial companies. So we are not afraid of building and creating the first clients, based on people connected to us, and my own experience and the experience of our General Partners.

We are not competing against other venture firms or private equity firms, in the following way. Usually, we will not lead in later rounds of funding. We will start companies from scratch. Or we will come in after a year, or even in later rounds. But somebody else has to lead the round. So our model is to get involved after we have been invited in by the lead investor, so we don't compete with them. We aim to be sought after as a value-added part of a syndicate that can make companies successful.

So we look for investment partners who are interested in deals that fit our thesis. And that allows us to be very, very choosy. Typically people have found that it helps them limit the deal if they can bring us to the table to have us involved with the CEO. In the case of CityBlock Health, there were three different teams competing for the deal. But we were part of the syndicate, no matter who won the deal, because the CEO wanted us, and all the firms knew us.

So we have a little bit of a different approach in that we don't view ourselves as competing. In fact the point is that our fund, whatever size it is, have enough capital to invest in this space. Our goal is not to invest $100 million or even $500 million in this space. Our goal is to attract $5 billion of capital into this space, and to use our involvement to leverage other firms.

Kevin Pereau: Are there "areas of pain" in health care that you are excited about working on?

Andy Slavitt: As it gets to national issues, mental health is huge. Kidney care and a new way of doing dialysis will be huge. Autism will be an enormous opportunity. Healthy baby delivery, and first 1,000 days of care, and family planning.

Did you know that Medicaid pays for half the baby deliveries in the country? And it pays for the care for half the kids in the country. So Medicaid should evolve toward thinking about smarter ways of doing what it does, and realize that it's about moving a lot of things together. I see great importance in the cycle of supporting individuals in planning pregnancies, in improving healthy baby delivery, and focusing on the first 100 days of care. Someday, somebody will turn that into a bundled payment which will look not only at maternity care and delivery, but also the outcome of having a healthy three-year-old.

Some of the things I am talking about are obviously nearer term opportunities, and some of them are bigger-picture, more challenging opportunities. But those are all kind of key spaces.

And of course, there is a lot to be done in primary care, primary care coordination, especially for senior, frail, elderly programs. Those are areas that are seeing a little more activity, but there should be a lot more.

We are very focused on helping to build businesses that focus in vulnerable populations, and helping to build those markets. Any company that views those populations as part of what challenges them to be in health care, that's an interesting place for us to invest and work. And of course there is a lot in primary care and care coordination. There has been a lot of activity in those spaces historically, but there needs to be a lot more.

I think that the thing we have to be careful of, is that we don't concentrate solely on digital technologies that focus only on the population that already has high-speed Internet access, smartphones, and other significant advantages, like the freedom and resources to limit how many hours they work.

Kevin Pereau: So it again comes back to that working woman who lives two bus rides distant from her dialysis center, right?

Andy Slavitt: Yes. And another obstacle is that we tend to think about health care as averages. How many people can we get to increase their blood pressure compliance, for example? How many more people can we get to take their medication? As opposed to worrying about the variations. Because our biggest problems in health care are caused by the variations, not by the averages.

Sixty percent of us do fairly well in managing our health care. When we create a health care system that only serves those of us who are in the high end, we make the overall situation worse and worse. So the averages move up only a little bit because we only reach the highly capable people.

When the underserved population continues to be underserved, we wonder why the whole system doesn't get any better. If we have people in health care that are not focused on the whole community base that we need to serve, they are going to find themselves frustrated. They're making little bits and pieces of the equation better, but nothing fundamentally changes.

MAKING A REAL DIFFERENCE

The kind of consumer Andy Slavitt describes - the one who has a hard time getting to a dialysis center, who has little access to good local food or services, and whose health is in genuine peril - is not someone he is making up. Consumers like that are all too real. We can provide care for them at home, or see them in emergency rooms.

They might not know it, but the improvements in the care they receive are due to the efforts that health care visionaries like Andy Slavitt are making on their behalf.

EXPERT OPINION: ABNER MASON, CEO, CONSEJOSANO

Every year, new studies show that there are first and second-class citizens in the world of health care. Women, especially women of color, receive a quality of care that lags behind. So do people who reside in inner cities and in remote locations far from metro centers.

And then there is the issue of language. America's Hispanic residents especially have a difficult time interacting with physicians and other caregivers who do not speak Spanish or understand the nuances of their culture.

But what if technology and telehealth could roll away many of those obstacles?

Abner Mason is CEO of ConsejoSano, a Los Angeles -based company committed to making high-quality health care accessible to underserved populations in the United States. At ConsejoSano, which means "healthy advice" in Spanish, Abner and his team are now developing a platform that utilizes telemedicine, SMS text messaging and other technologies to serve America's Latino population, including undocumented Spanish speakers.

Let's hear what Abner Mason told me in a recent conversation about his remarkable experiences and company.

Kevin Pereau: Can you tell me a little bit about your background and how you came to arrive at this very noble mission?

Abner Mason: I'm from North Carolina originally. I went to prep school in Massachusetts, then to Harvard. After Harvard, my first job was at Bain Consulting. From the very beginning, I was working around health care. One client was a pharmaceutical company, and another was a hospital health system.

Then after Bain, I got involved in the state government of Massachusetts, for what I thought would be a very short period. When William Weld ran for governor, I supported him. He won and said to me, "Come work in the state government for two years and bring some fresh ideas."

I accepted that invitation, but I ended up staying for more like 14 or 15 years. I took lots of different roles, but ultimately, I was Chief Policy Advisor to Bill Weld, and then for Paul Cellucci and Jane Swift, the two governors who succeeded Weld. I left Bain just before a fellow named Mitt Romney became governor. I knew him at Bain, where he started Bain Capital. We did a lot of the lead-up to what would become known as Romney Care, which was the first such program in the country that offered universal coverage.

Then while I was still working as the Chief Policy Advisor for the State of Massachusetts, I got involved with the Presidential Advisory Council on HIV/AIDS under George W. Bush. It was a voluntary position to advise the president, the Secretary of State, and the Secretary of Health and Human Services.

AIDS at the time was a growing crisis, both domestically and internationally, and something had to be done. And I just fell in love with that work. It was an example of U.S. leadership in the world. After I joined the Presidential Advisory Council, I was appointed Chairman of the International Sub-Committee.

My focus there, in the early years of that administration, was to help conceptualize and implement the Presidential Emergency Plan for AIDS Relief, which became the largest humanitarian health effort ever in terms of money spent. President Bush announced it in his 2003 State of the Union, a $3 billion plan to get anti-retroviral therapy to people in about 15 countries, mostly in Africa.

At the time, the mortality rates for HIV were just astounding. Two million people were dying a year, mostly in Africa, but also elsewhere. The U.S. and other countries in the developed world had anti-retroviral therapy, and people in those countries were living longer with the disease. But that wasn't happening in Africa, so the U.S. took an incredible leadership role in terms of delivering anti-retroviral therapy drugs to people.

The view was that if we could put a man on the moon, we could get anti-retroviral therapy to people in the furthest reaches of Africa, in remote places without running water, without electricity, without refrigeration for medicines. It was very complicated. Health care workers rode motorcycles to go out into the bush in Africa, because there were no real roads. I spent time in Uganda, seeing them ride out in the morning to visit people in remote areas. The infant mortality rates were devastating, close to 100% for people with HIV. But we were able to slowly bring those numbers down by something like 80%. It was astounding.

So I decided to start a nonprofit, the AIDS Accessibility Project, and we focused on HIV work. Much of the U.S. government funding was going to Africa, for good reasons that I supported. But there was no new funding for Latin America, so I worked with corporations and started the first business council on HIV/AIDS in Mexico. We more than matched government money with private-sector money.

We started to challenge the biggest problem in Mexico, which was discrimination around HIV. People were being fired or not hired because of their HIV status or presumed status. If an employer thought people were gay, they wouldn't hire them because they might have HIV. It was crazy.

We got the support of the Secretary of Health in Mexico and the U.S. ambassador, and we ended up doing some really good work, getting companies to change their policies. It was a great experience. I learned a lot there about the role that employers can play. But values matter. It is not just about having the ability to do the right thing, you also must have values that you stand by and live by, that *compel* you to do the right thing.

Keven Pereau: This is pretty astounding. What did you do next?

Abner Mason: The experience of running a nonprofit led me to start a for-profit. The companies we were working with in Mexico said that they loved what we were doing with AIDS, but that they wanted to bring people better care for other diseases too. So my partners and I created a for-profit called the Workplace Wellness Council of Mexico, which has now become the leading corporate wellness company in

Mexico. It's a consulting firm that helps companies design and implement employee wellness programs. It has more than 100 corporate clients now and is working with them to improve health across the board, not just HIV, but generally.

Keven Pereau: Okay, you have gotten us as far as Mexico in your story. How did you get back to the States and start ConsejoSano?

Abner Mason: I saw that there were a lot of solutions that were happening in Mexico for Spanish speakers. And there was an immense need for similar solutions here. You know, Mexico's population is something like 119 million, and we have about 60 million Hispanics in the U.S., most of them Mexican.

I saw that telehealth was growing really fast in Mexico. But telehealth was not growing at all in the U.S. or growing very slowly. So I had the idea to create ConsejoSano. My initial goal was to focus on Spanish-speaking people in the U.S. and to try to improve their health with the application of telehealth technologies.

Kevin Pereau: Over the years, one charge I've heard leveled against George W. Bush was that he turned his back on the AIDS crisis in Africa. But what you said earlier completely goes against that.

Abner Mason: Nothing could be more untrue. He fought hard to get funding for AIDS relief in Africa. There were members of Congress who were saying, "We have seniors in my district who need help, and you're going to give medicine to people out in the bush in Africa?" You could imagine it was not the easiest thing to do, to get Congress to step up to do the right thing. But it was a health crisis that could literally have spun out of control.

Historically, as we get some distance from George W. Bush's administration, people will see that one of the areas where he really showed extraordinary leadership was in health and HIV. It is one place where I think history will see him as a great leader, and a moral leader.

Kevin Pereau: Please tell us more about ConsejoSano.

Abner Mason: ConsejoSano is a startup. We are venture-backed, and we have a group of committed investors that includes 7wire

Ventures, Tufts Health Ventures, TOTAL Impact Capital, Wanxiang Healthcare Investments, Acumen, Oxeon Partners and Impact Engine. Because we are venture-backed, we are under pressure to perform because we are for-profit. We need to grow the business, and we will have an impact on health, but also hopefully we will reward investors who believed in us.

As we have grown the business, we see that Medicaid offers a huge opportunity for us to do a lot of good. We are also working with Medicare Advantage. But I envision that we will be a thought leader when it comes to Medicaid. We believe that Medicaid is crucial to the country's overall health. So many people are on Medicaid, it is the largest health program in the country. Someone has said, and it's probably true, that the number of people who are on Medicaid is equal to the population of France. It is a huge program, and it's growing, and it's crucially important. That doesn't mean it's perfect and it shouldn't be improved, but it is important to millions of Americans.

One of the things we need to do is figure out how to move innovation faster within the Medicaid program. Government is notoriously slow, with all kinds of regulation. Health care is also notoriously slow to change. If you combine them, you get almost glacial slowness. And then you add to that the fact that Medicaid is a federal and state program – actually 50 different programs. There is no one set of rules.

Now, that could be turned into an opportunity. It could be possible, for example, to let 50 different states try 50 different things. They could be laboratories for experimentation. I'm hopeful, but it hasn't yet happened nearly the way it should.

Kevin Pereau: Why do you think that certain populations in America are so underserved when it comes to health care?

Abner Mason: As is the case with a lot of things in our society, it's not easy being poor. That's the reality. And health care is expensive. If you are wealthy, you can pay for it yourself. If you qualify for Medicaid, fantastic, you have some support. But there are challenges for the people in that middle category. They're not poor enough to qualify for Medicaid, but not wealthy enough to just buy health care on the open market. So, they suffer.

And then of course, there are the undocumented. We have a lot of people in the country who don't qualify for Medicaid because they're not citizens, but most are still low income.

Kevin Pereau: How do you go about providing better care for the poor and undocumented?

Abner Mason: I think it will be a combination of technologies like telemedicine and various programs.. I'm hopeful, but it's not going to happen in the short term. California has moved to start to provide some health care for undocumented children for example, and pregnant mothers. Even if they don't have papers, they can get care. California is an example of a state that is trying to do some of that now.

Kevin Pereau: What else can you tell us about telemedicine? Does it really have the potential to help vast groups of underserved people?

Abner Mason: I think so. And it is, but not yet to the extent that it could. I was really impressed by what I had seen on telehealth in Mexico. It was growing fast and interestingly, in Mexico, it was targeted toward the growing middle-class. In Mexico, wealthy people have a lot of options. A lot of them come to the U.S., to Europe or to England for health care. But if you're poor in Mexico, you have to rely on the government, and it's not great. The government has a universal health care program that was instituted by Vicente Fox. But it's government care, and as you might expect, it's not always the highest quality and there are long waits.

There is a growing middle class in Mexico, because the economy is growing. Middle-class Mexicans wanted something better than the government programs and they were looking for solutions. So Salud Interactiva, a telehealth platform, developed a solution.

I got to know that company very well—it's a leading telehealth company in Mexico. A few years ago, it had 5 million subscribers, all getting high-quality telehealth services. McKinsey did a study and recognized it as one of the global best practices for telehealth in the world.

I got to know the people in that company and I partnered with them. We had discussions about offering their service to Spanish speakers in the U.S., where there are not enough doctors who speak Spanish. So I put together a partnership with Salud Interactiva in Mexico that would allow people in the U.S., Spanish speakers, to just tap their mobile phones and get medical counseling. It would connect customers to native Spanish-speaking physicians 24 hours a day, seven days a week. You couldn't do a full consultation, because the doctors are obviously in Mexico and they're not licensed in the U.S., but they could give advice.

There are 60 million Hispanic people in America, and about 20 million of them speak Spanish only. Many of them are low income and many of them are undocumented and can't qualify for Medicaid.

There are obstacles. With government programs, for example Medicaid, you cannot use offshore resources. But in my opinion, there's a huge potential to significantly improve access to high-quality health care for Spanish speakers in the U.S. if we could just allow them to use this service that's just across the border, a fantastic service that middle-class Mexicans are using. Plus, because of the price differentials of our two countries, we could make it extremely affordable for lower income people in the U.S. But you can't do it because of various, what I would call, outdated regulations. It's another example of a huge missed opportunity to help people. But we are working on that.

Just think about it. If you are a Spanish speaker in, say, San Antonio, you could talk to a doctor in Mexico who is only a short distance away, someone who speaks your language and understands your culture. But you are prevented from doing that. You have to speak with an English-speaking physician who is somewhere in America, possibly very far away. So this web of regulations hasn't quite caught up to the reality of telemedicine. Your mobile phone allows you to talk to anyone, so why shouldn't you be allowed to talk to a doctor in Mexico? A physician of your choice?

Kevin Pereau: I will be thrilled to see where you go now with your company, and equally thrilled to see how much good you will do in the world.

EXPERT OPINION: DR. RAY COSTANTINI, CO-FOUNDER AND CEO, BRIGHT.MD

For many people, a trip to see a physician is a lot like a trip to the Department of Motor Vehicles. In both cases, they are going to wait a long time, fill out a lot of forms, and be frustrated because they forgot a piece of documentation. And of course, they know they will be grumpy because they lost a half day of work.

As you might have noticed, computers have improved the end-user experience at most DMVs. Yet for many patients, a trip to see a doctor is still an experience that has not improved much at all. Waits are long and frustrating, the amount of time spent with the doctor is too short.

Yet a number of young entrepreneurs are developing technologies and systems that will soon change all that.

Ray Costantini is co-founder and CEO of Bright.md, a company that was founded to provide automation support to assist with primary care provider workflows and help them provide more thorough and evidence-based diagnosis and treatment.

Bright.md is developing a new range of innovative technologies that include its SmartExam, which the company describes as, "your virtual physician, cuts the cost of visits by 80% - while providing high-quality care your patients will love."

To use SmartExam, patients go online and use its simple interface to provide information on their medical histories, current ailments, as well as information on their insurance, preferred pharmacies, and more. In their offices, physicians receive an immediate alert that a patient report has been filed. All information is automatically entered into that physician's system. He or she can immediately issue prescription orders, schedule an appointment and take other appropriate steps - all in a matter of a few minutes. That is only one example of the systems that Bright.md is building to improve the care experience, both for patients and their care providers.

Bright.md was named the 2015 OEN Launch Stage Company of the year, a 2016 Gartner Cool Vendor in Healthcare Delivery, and the HiMSS Venture+ 2016 Audience Choice winner. The company also received the 2017 Healthcare Innovation Award and was chosen as one of the most admired health care companies in the Northwest.

Ray's unique pedigree and background have equipped him to have a highly informed perspective on what his happening in health care. He was first an entrepreneur: "My father and grandfather were entrepreneurs," he tells us, "And I actually started my first company, which was in real estate, after I graduated from high school. Then I decided to go to medical school; at that time, my father gave me a look that pretty much said, `Why?' But then I came back and became an entrepreneur again."

Part of the Bright.md mission is to "delight" both patients and their providers while improving clinical outcomes, dramatically lowering health care costs, and developing opportunities for health care providers and businesses to discover opportunities for new revenues.

Those are lofty goals to achieve in today's health care landscape. We thank Ray for contributing his insights to *The Digital Health Revolution*.

We know the readers of this book will find this excerpt from my recent conversation with Ray to be very stimulating and, dare we say, revelatory?

Kevin Pereau: What was the thinking behind Bright.md?

Ray Costantini: There are a lot of layers to that question. I'm a big believer that the most important problems to solve are the ones that are so big and so ubiquitous that people have forgotten that they are there.

I think that Bright.md falls into the category of companies that are asking those questions. On the surface, we solve problems like: how do we help patients get pinged more quickly by their providers; how can we help them get care more quickly or conveniently; and, how can we make care more affordable?

And then, maybe one level up, there is the experience factor. How can we make the experience of health care if not more enjoyable, at least less unenjoyable? Health care is something that everybody needs, but nobody really wants it.

But at its core, the ubiquitous big, hairy, issue is that providers have what patients need, but it is just too difficult and inconvenient to deliver it, because they are spending so much time on menial stuff.

Today's care providers are just too rushed, and our company was created in part to address that problem. The old country doctor model, where patients knew their clinician, who in turn had all the time in the world to provide care, was a beautiful one. But it isn't going to work in today's society. Today, physicians spend so much of their time dealing with an administrative burden. They're filling out forms, online forms at this point, filling out paperwork and documentation, doing order entry, and creating billing files. Those activities are nowhere near the top of the list of most important things that they should be doing.

Kevin Pereau: And the net effect on the whole health care system is . . .

Ray Costantini: We've got a huge supply/demand mismatch. Taking our analysis up a step, we've got a problem with elastic supplier capacity. Under the current circumstances, we can't really grow how many patients the provider can see. We can't fix that problem by using traditional approaches.

In some ways, it's like the agricultural work evolution over the last century, where we had an explosion of population and our farmers needed to be able to feed everyone. The challenge was, how do you feed a gigantic and growing population of people with farmers out there wielding shovels and picks? Instead of thinking farmers could meet the demand with those tools, we equipped them with a combine or a tractor. That made it easier for them to do more with the limited time and capacity that they had.

Today, we're facing something similar in health care. As our population grows and ages, our demands on our health care system are growing as well, at an exponential rate. If we try to keep solving health care needs with picks and shovels, with the tools that we've always tried to use, we can never figure out how to work smarter.

The platform we have built at Bright.md is designed to help providers work smarter. It's there to give them the support and tools that they need. That's why we think of ourselves not as a health company or a care company, but as a care automation company.

Kevin Pereau: So, if you boil it all down?

Ray Costantini: One job is to make physicians' lives better by empowering them with the tools they need to better serve their patients' needs. And that's where the concept of care automation comes in.

How can we gather the information that's needed, how do we do that in a delightful, and highly functional, way? How do we synthesize that information and make it is as useful and actionable as possible? And how do we enable care providers to take action quickly and easily? We are in the process of figuring out those pieces.

Kevin Pereau: Can I ask you to explain what some of those "pieces" are?

Ray Costantini: Let me take a step back and discuss care delivery from the patient's perspective. Care falls into one of two categories. You can either decide to dissect everything down to its smallest pieces, or you can look at categories. And I definitely fall into that second bucket - I'm a lumper rather than a splitter. And so, in my mind, there are two big buckets of care from a patient perspective. I would call them pull-based care and push-based care.

Pull-based care is care that patients actively want and demand - the category of care that patients are actively pulling to themselves. If

I've got a headache, for instance, I am looking for immediate help for that, so I pull that care to me.

Push-based care is different. It is care that the providers need to push to the patient, because he or she might not even know to ask for it. One example is care for high blood pressure. Care providers need to push care for that problem to the consumers, who might not even know they are suffering from it. Do you know what high blood pressure feels like? It feels like nothing . . . nothing! And the same push-based model applies to other conditions, like high cholesterol and, for the most part, diabetes. Those are clearly push-based conditions, where caregivers need to help patients recognize that they have a need and to help them get care.

Both kinds of care are important. But in creating our company, we made the choice to start in the pull-based care model. That has been the best place to be able to provide care solutions, because patients were already asking for them. So, we built a whole set of tools that help automate the care of more than 370 different conditions. Our tools allow patients to request care from their own providers, or to use our SmartExam. From there, we built tools that allow patients to complete their medical data - in effect, to interview themselves. Our system aggregates patient data in a highly usable format so that providers can quickly and easily understand what is going on with patients, so they can order prescriptions, follow up and monitor what is taking place.

The goal is to build all the tools needed to support each of the tasks that providers need to perform, by aggregating them into a painless workflow. That makes life easier for everyone.

The individual tools are not always particularly interesting. Some are more innovative and exciting than others are, obviously. Some of them are more unique, like our ability to allow patients to effectively interview themselves, or our ability to convert multi-source data into an aggregated, chart-ready note. But the exciting part is how we put those pieces together to support the providers' workflow, which in the end makes patients' lives easier. That's really what this is about.

In the end, the people in my company are the only ones who care deeply about the features we have built. Our real job is to think about the value that those features create. What really matters isn't the tools that we build, it's about how we can apply them to create value for patients and providers and for the health system overall.

Kevin Pereau: When did you start to think about health technology from that perspective - from the end-user's perspective first?

Ray Costantini: Let me tell you one of my favorite stories. Here in my office, people are tired of hearing me tell this one, but here we go. The story is about something that happened when I was in medical school and was doing my first surgical rotation.

It was at a VA hospital. The surgeon who was supervising the rotation was a tough, rough physician and teacher. He was a stereotypical surgeon in that way. He was fond of hitting the med students with all kinds of tough questions. I was prepped. I knew he was going to give me hard questions too. Then the time came when I was stitching up a patient after doing my very first surgery. Just as I was closing up, he said to me, "What are the two most important parts of any patient surgery, no matter what kind of surgery it is?"

I didn't have any idea. I had never encountered that question in any textbooks. After he let me think for a few seconds, he said, "The two most important things are the scar and the bill, because those are the only two things the patient sees, so you better make them both look good."

It was a tongue-in-cheek observation perhaps, but that experience has stayed with me. Patients don't know what happens inside of them during surgery. They assume that their surgery was done well. They only measure success and satisfaction on the things that they can see, and that is also true whether you are discussing a software company, an insurance company, or any other kind of enterprise. What matters most to consumers are the pieces that they experience. If what they see is satisfactory, they assume that the rest of what you have done for them has worked right - unless they can prove otherwise.

Kevin Pereau: Looking ahead, how do you think contributions by a company like yours will change how end-users perceive the care they are getting? Or addressing a larger issue, how will the world of health care change through the application of technology?

Ray Costantini: It's hardly a profound observation, but I think health care is changing faster now than ever before. And I think that there's a terrible misconception out there, when people say that everyone in health care is tech-averse. I think that's absolutely false.

Did you know that physicians were the leaders in starting to use smartphones? They were the first group to exceed 90% penetration. Physicians use new technology every single day, they love their newest tech toys.

In the past, the primary quantifiable goal of many players in health care was not to keep people healthy, but to fill hospital beds. That is certainly changing. If you look at the billboards on the roads near where I live, you will see that the messaging is changing. It is less, "come to see us when you are sick," and more, "we are here to help you stay healthy."

And as we move from the fee-for-service world into a value-based care world, that's a fundamental realignment of interests, because health care is discovering and addressing what patients want from them. And that, I believe, is huge.

EXPERT OPINION: LEAH SPARKS, CEO & FOUNDER, WILDFLOWER HEALTH

Meeting the needs of the underserved is exceptionally challenging. Because telehealth and population-health technologies help us manage diverse and dispersed demographics, you would think tools for managing things as straightforward as women's and family health would be robust and plentiful. Yet family health has remained stuck in a gap. Why is that? Often, because the wrong people are making the decisions. In once glimpse, a friend of mine recently forwarded me a photo of healthcare policy being set in Washington, DC. There was not a single woman in the frame! Fortunately, in the wider world of healthcare innovation, that is not the case.

More often than not, it is a woman who is selecting the family healthcare plan, scheduling the doctor appointments, reconciling the co-pays and generally doing all the heavy lifting. It starts early, from pre-natal to caring for young children, on through to adulthood. Understanding a woman's perspective as the family healthcare point person is a huge advantage. Combine this with many women's personal experience and passion for improving things, and you get Wild Flower Health. Meet Leah Sparks.

In 2012, Leah was pregnant with her first child. Navigating regular doctor visits, she recognized that what should have been a time of joy was often tempered by uncertainty about health plan benefits, hospital services and appointment schedules. She realized pregnancy could be a powerful context for teaching women how to use the health care system in a smart way.

So Leah started Wildflower Health. In the years since, Wildflower Health has taken a transformative role in helping families connect to quality care.

Here are some excerpts from a recent conversation I had with Leah about her, her company, about the state of care today, and about the future.

Kevin Pereau: What was the impetus behind starting Wildflower Health?

Leah Sparks: I started Wildflower Health in the summer of 2012 when I was five months pregnant with my first child. At the time, I had worked in health care for nearly 10 years and thought I knew how it worked. It wasn't until I decided to start a family that I actually had to navigate the system firsthand. Suffice it to day, the experience left a lot to be desired. I didn't know which hospital was covered by my health plan and was confused by my medical benefits. The real kicker was when I received a mailed letter from my health plan telling me I was a high-risk pregnancy, just because I was over 35, and to please call a 1-800 number to talk about it.

Meanwhile, I was using all these slick consumer mobile apps to help me with everything from my baby registry to picking out a stroller. I thought that if we could bring that type of consumer experience to health care, we could not only make the experience about 10 times better for new families, but we could make a dent on quality and outcomes in maternal-child health. And once I dug in further, I learned that the United States lags far behind other OECD (Organization for Economic Cooperation and Development) countries on quality measures in maternal health, which translates into tens of billions of excess costs to the health care system. The scale of the problem from a cost perspective gave me the confidence that the system would pay for the type of solution that Wildflower Health could provide.

Kevin Pereau: You have a family health product designed from the ground up for consumers, yet the business model used to accelerate your access to market is B2B2C. What lessons can you share about getting consumer products to market through complex relationships with payers, providers and employers?

Leah Sparks: Today, we sell to health plans, hospitals and self-funded employers – all of which can be complex sales cycles to navigate. I'm very proud of the fact that we work with some of the largest health plans in the country, with contracts that cover nearly 50 million lives. Back in 2014, our contracts only covered one million, so we've experienced tremendous growth. Our average sales cycle is about 8 months, which is actually very good for health care enterprise software. The health care industry as a whole is not particularly consumer-savvy, which can make it hard to push through innovation and to implement solutions that resonate with consumers. Added to that are the challenges with security and privacy requirements, bureaucratic approval processes, and just managing to budget seasons with each client.

However, the good thing about health care is that the vast majority of the people working in this industry are doing so because they really care about making a difference in people's lives. And over and over again, we've found individuals in these behemoth, complex organizations who share our passion for improving the health and happiness of women, children and their families, and are willing to work with us to knock down the obstacles to innovation in health care. Our sales team's mantra is "start with the why" – always remind our buyer why we are here, why we do this, why it matters to them and the people they serve. Ultimately, that's not only the best way to ignite the type of passion that helps us get sales closed, but it also helps us build long-term relationships with our client partners and drive ongoing innovation with them.

Kevin Pereau: Let's talk about smart money for a second. Rock Health tells us in their 1H/2018 Investment Report we are in the midst of another record-shattering year for health care investing. How important is it for companies like Wildflower Health to work with investors capable of helping them navigate the challenges of bringing innovation to market through such complex channels?

Leah Sparks: I'm a first-time founder and I was a bit naïve to venture fundraising when we first started the company. Since that time, I've learned how crucial it is for companies to match their business plan to the right investors. There are a couple of key aspects that are important to get right in that match. First, if you are an enterprise health care company, it's essential to have investors with deep experience in health care, and better yet if they can help connect you with customers. The health care sales cycle is a slog and you need investors who will have the patience to stick with you as you figure it out. Second, the majority of digital health companies who have successful exits will get taken out by acquisition rather than IPO. Many of those acquisitions will be modest. Depending on the size of the exit you think your company can achieve, it's important not to take too much capital or raise from funds that need $1B+ exits to meet their own internal return thresholds. I have a mix of financial and strategic investors who are all well-aligned with our business strategy and commercial plan. As a bonus, my Board members are terrific advisors to me and my team and great people to be around. That is a huge and under-appreciated key to one's sustainability as a venture-backed CEO.

Kevin Pereau: Let's focus for a second on who actually consumes health care. You have coined a phrase I love...the Chief Health Officer. Describe the typical CHO -- how can we make health care a more consumer-friendly experience and what problems do we need to solve for her?

Leah Sparks: The data is very clear – women make the vast majority of health care decisions, not just for themselves, but for their entire families. Often, this journey as the "Chief Health Officer of the Home" begins with parenthood, but it can also start with caring for a partner, an aging parent, or themselves as an individual. And certainly some of the "CHOs of the Home" are men. But whoever that person is for the family, we aspire to make the CHO's life easier through personalized and elegant connections to health care.

What we hear from our users is that they are struggling to keep track of all the health needs for their family and plug into the different silos of health care when they need it. In a user's words, I would sum it up as – "I'm tired of having to go to so many different places to get what I need for my family's health." She has one portal for her kid's vaccines, another one for her employer's benefits, and maybe somewhere else she has to go to to make appointments for herself. On top of that, when she switches employers and her health plan changes, she has to start the process of "finding stuff" all over again.

Our vision at Wildflower is to reduce that frustration and fragmentation through an elegant technology experience that follows the user from pregnancy through pediatrics and family caregiving, and that connects her to plans and providers all along the way. Given that we have a huge footprint of health plan and employer clients, as well as a growing base of provider clients, we believe we are well-suited to execute on that vision create a truly game-changing window into health care for the Chief Health Officer of the Home.

Kevin Pereau: Where do you see health care in 10 years?

Leah Sparks: Last weekend my youngest son's hurt finger looked like it might be getting infected. I called my pediatrician's nurse advice line and left a message. I got a call back about 20 minutes later and she sent us to urgent care. We drove to urgent care and waited for about an hour before having a 10-minute visit with the doctor. Then we had to go to CVS, hand the pharmacist our handwritten prescription, and wait another 30 minutes to get it. All of this with a three-year-old boy on a Saturday evening.

Ten years from now, a parent in that situation will use a digital symptom checker with intelligent logic, which will then connect the family to an on-demand video visit to take a look at that infected finger. The telemedicine clinician will then send the prescription electronically to a pharmacy that will deliver the medication to the family's home within two hours. Three-year old boy gets to go back to his Legos and his mom gets that much-needed glass of wine. Win-win.

AFTERWORD - A DAY IN THE LIFE

Like most people, I live for the weekend. The sun rising over Mount Diablo wakes me as it shines through the back sliding door in our bedroom. You can't beat California weather from about April – October. It is warm, dry and moderate. The coffee is on and I can hear my wife moving about the kitchen. Beth is what you would call "Type A." She gets more done before breakfast than most of us do in a week. You don't get to lay about, even on the weekend, when you are married to an overachiever. I get a move on. We have places to go and things to do.

I get out of bed and step on our Withings scale to check my weight, BMI and body fat. Next, I strap on my blood pressure monitor and take a reading. The entire process takes about two minutes and is baked into my daily routine. My weight is holding steady at 180 pounds. It is not my ideal weight and I have some work to do to get down to my goal of 172. BMI and body fat are also a little high, but in an acceptable range. I can remember a time when I had no idea what Body Mass Index even meant, let alone what was normal. My blood pressure is 123/78. I lost my father to a heart attack and he had chronically high blood pressure, so I have learned to track mine. I have been fortunate. So far, good nutrition and plenty of exercise have been effective in keeping it in line.

I like to tell a simple story whenever I keynote health care events or participate on panel discussions about why this routine is so important. My niece was getting married and decided to do a destination wedding in Cabo San Lucas, Mexico. My wife, wanting to look good in her bikini, bought some fat-burning pills to help with that pesky winter weight gain we all experience. Who could blame her?

I quietly started taking them myself. Hey, men like to look good too! Over about a two- to three-week period, I lost a whopping 12 pounds, getting below my target goal of 170. I hadn't looked that good since university! I couldn't wait to step on my scale every morning. It was when I checked my blood pressure that I got my wake-up call. I was shocked to see my readings go from a consistent 123/78 range to a whopping 165/105!!!!

I was looking great but inviting a cardiac event with how I was doing it! For a man whose father had a history of high blood pressure and died of a heart attack, my taking fat-burning pills was probably not the most sensible way to achieve my weight goal. How would I have known if I wasn't tracking my biometrics daily? Furthermore, how would I know what is an acceptable range for a man of my age without getting weekly tips, tricks and traps for how to better manage my health? Later in the year my physician Dr. Sauers would isolate this period and ask me, what the heck happened?

If I wasn't clear before on the dangers of maintaining my weight using tactics like the pills, I am now thankful to my doctor because he can connect to my lifestyle activities and nutrition choices. He remains my very best resource, confidant and coach.

The smell of the coffee finally lures me into the kitchen, where I join my wife who has prepared us a bowl of fresh fruits and yogurt. I like to sprinkle oats on mine. A little side of berries from our garden, and I am all set.

It is a swimming day for me, so I grab my backpack and head to the gym. I do the three-mile walk to the train station without even breaking a sweat. The train station is right across the street from our gym. On days when I am working in San Francisco, I ride my bike to the gym and store it there while I am in the city. There was a time when walking more than two blocks was daunting. Now, it is a breeze. I get to the gym, do my stretching, and head outside to wait for a lap lane. It isn't a busy day, so I get a lane to myself and swim for about an hour. I am not the fastest swimmer there, but I average right around a mile. After a steam and shower, I check into the restaurant and order a house salad with a cup of soup. The walk back home always takes a little longer because it is uphill, and I am a bit knackered. But it is a great way for me to decompress and clear my head of all the things I have going on in my life. I have energy and enthusiasm.

It wasn't that long ago that I weighed 238 pounds, and barely had energy to do much of anything. Obesity strikes practically overnight when you are an adult. Losing weight takes a lot longer. Today, I am residing at the corner of happy and healthy. I marvel at how my life has truly come full circle. As a child growing up in Vermont, I was constantly on the go, eating healthy foods and looking forward to life events, I had enthusiasm, curiosity and energy for just about everything.

It is amazing how re-embracing that life style has made me feel like a kid again. Whenever my wife and I go anywhere, our first instinct is to walk there. Whether it is picking up the dry cleaning or stopping by the market for our fresh produce and dinner for the evening, it all adds up.

My workmates often marvel at how a man my age can average 5-8 miles of walking every day. It isn't hard, but you have to start somewhere. I will share a little secret with you: the more you exercise, the better you feel. When you throw good nutrition into that mix, you develop an addiction for a healthy lifestyle that makes you feel good, look good and have confidence and enthusiasm for life. There is no turning back.

I am amazed at the tools we now have available to us. Many are free. Every day, I open my Apple Health app and check my activity, nutrition, sleep and mindfulness. I am counting steps, exercise minutes, total time spent standing, you name it, there is an app for it! I am not here to recommend one over another or to promote my health scoring company. I am here to tell you that the more you explore, the better you understand how lifestyle choices affect you in an incremental way, a longitudinal way.

Not sure whether Fooducate or NUMi is better for you? Try them both! What have you got to lose except weight and your waist size? Not sure how staying spiritually centered or getting enough sleep affects your health? Find out by downloading Zen Relax or Beddit. Most of these apps are free, they all hit your Apple dashboard and they all help you collect, analyze and predict health outcomes by better connecting you to the stakeholders who keep us healthy. Throughout much of our lives, we have always thought of our biggest stakeholders in our health as our doctors. Increasingly, we are connecting to a broader set of resources that can include nutritionists, fitness instructors, psychologists, life coaches, hospice and more.

If Digital Health 1.0 was all about proving we will engage and Digital Health 2.0 was analyzing all that data we have been collecting, then Digital Health 3.0 will most certainly be about connecting that data in a consumable way to our health care value chain partners.

At every step, consumers have led the way. Even in the early days of Digital Health when the providers and payers were deriving much of the benefit, Fitbit and others were quietly focusing on consumers and discovering that engaging them between trips to the hospital could ultimately mean fewer trips to the hospital. Very soon, your annual checkup will be a thing of the past, because you will be so connected to your doctor that he or she will be able to tell you that things are fine.

Doctors will turn their attention toward keeping us healthy, not simply caring for us when we aren't. They will look for better ways to connect and influence our behavior by leveraging digital health assets that connect to powerful Artificial Intelligence engines like Watson. All you will need is your phone and a little aptitude and inclination. I never believed or accepted for a single minute when I heard conventional wisdom held that we won't do what we need to do to be healthy. Hogwash!

So, what about that epiphany moment I described at the start of this book - the one I had while listening to the speakers at that health care event? What is the secret sauce? What keeps us healthy and out of the hospitals? We are the key. Our daily lifestyle choices matter. They are everything. We can't cure health care with an ACA (the Affordable Care Act) or an ACO (Accountable Care Organization) alone. It requires our participation. I keep waiting for some brave politician to step up and encourage us to channel our inner John F. Kennedy, "Ask not what your country can do for you. Ask what you can do for your country." In health care, that might translate to, "Ask not what your country can do to make you healthy. Ask what you can do to make your country healthy."

If you really want to do your part in fixing (your) health care, I leave you with three things to remember:

1. Exercise,
2. Eat nutritious foods that are low in sugar,
3. Cut back on alcohol.

See you at the corner of happy and healthy.

INDEX

V

W

X

CONTRIBUTORS

Oron Afek is co-founder and CEO of Vim, a company that partners with medical providers and health systems to provide convenience, cost savings and a better care experience to patients.

Beth Andersen is a leading figure in the world of health innovation. She has held top executive positions at large insurance firms and advised start-ups. Among her other current positions, she serves on the board of directors at Wildflower Health.

John L. Brooks III is CEO of Healthcare Capital, LLC, a consulting company that advises companies that are developing advanced, disruptive health care solutions, especially for diabetes. John has co-founded seven life sciences companies, including Insulet, developers of the Omnipod® insulin-delivery device.

Dr. Ray Costantini is Co-Founder and CEO of Bright.md, a company that was founded to provide automation support to assist with primary care provider workflows and help them provide more thorough and evidence-based diagnosis and treatment.

Bill Daniels is a Certified Elite Personal Trainer who serves clients at Renaissance Club Sport, a top exercise club in Walnut Creek, California. He is a recognized leader in the application of wearable and other technologies for athletes.

Sean Duffy is Co-Founder and CEO of Omada Health, a company that he founded in 2012 to offer a new way to help people with prediabetes prevent the onset of the disease, and to help people who already have type 2 diabetes manage their disease more effectively. Today, 150,000 members are using Omada to lead healthier lives.

David C. Edelman is Chief Marketing Officer, Aetna. His current responsibilities include refocusing Aetna's marketing on the individual customer through the addition of new retail pharmacy facilities and services.

Brad Fluegel is former Senior Vice President, Chief Healthcare Commercial Market Development for the Walgreen Company, where he was responsible for all commercial health care activities, including sales and contracting, biopharma relationships, retail clinics, clinical affairs, new service development and market planning.

Michele Geraldi is Communications Specialist at Omada Health, a company that was founded in 2012 to offer a new way to help people with prediabetes prevent the onset of the disease, and to help people who already have type 2 diabetes manage their disease more effectively. Today, 150,000 members are using Omada to lead much more healthy lives.

Matthew Holt is Founder of Health 2.0, a company that he founded with Indu Subaiya in 2006. Its annual Health 2.0 conference brings together representatives from every important company on the digital health care landscape.

Dr. Rajiv Kumar is President and Chief Medical Officer of Virgin Pulse, which is part of Sir Richard Branson's Virgin Group.

Henry Loubet is CEO of Bohemia Health and a leading broker/consultant in the world of health care. He has negotiated agreements between insurers and companies that have brought health coverage to millions of consumers.

Dr. Mike Lovdal is a leading health care analyst who earned his MBA and Doctorate at Harvard Business School, then joined its faculty teaching corporate strategy. Today, he serves as an Adjunct Professor at the Columbia Business School, co-teaching a course on Innovative Models in Global Healthcare.

Richard Lungen is Founder and Managing Member at Leverage Health Solutions, which describes itself as, "the market leader in Healthcare

Growth Services focused on health care Strategic Business Development. We concentrate on delivering best-in-class solutions by way of our Portfolio Companies to the health care marketplace, with an emphasis on payers and health plans, based on unique and unparalleled industry experience."

Tom Martin is Senior Vice President and Chief Strategy and Information Officer at Evergreen Health, a two-hospital system in Kirkland, Washington, that is one of the largest and most innovative care providers in the Pacific Northwest. Tom is responsible for information technology, strategic planning and business and relationship development.

Abner Mason is CEO of ConsejoSano, a Los Angeles-based company committed to making high-quality health care accessible to under-served populations in the United States.

Penny Moore is a Partner with Commonwealth Health Advisors. She has devoted her career to providing leadership to growth-oriented companies that are transforming population health through innovation with digital technology. Most recently, Penny has led the commercialization for early stage digital health start-ups Kurbo Health and Better Therapeutics. As the Chief Revenue Officer for ShapeUp, a venture-backed company that pioneered Social Wellness, Penny's leadership created a high-quality book of global business that paved the path to a successful exit.. She has held executive leadership roles with large health plans such as Kaiser Permanente, UnitedHealthcare and Aetna.

Dr. Thomas Munger is Heart Rhythm Division Chairman at the Mayo Clinic in Rochester, Minnesota. He was awarded the Mayo Clinic's Cardiovascular Laureate Award in 2007, was named the Cardiovascular Division Teacher of the Year in 2002 and has won numerous commendations that reflect his contributions in education, research and patient care.

Christine Paige, PhD, is Senior Vice President, Marketing & Digital Services, at Kaiser Permanente. She leads national marketing, digital services, and consumer strategy for Kaiser Permanente — the nation's largest not-for-profit health system serving over 12 million members and more than 125,000 employer groups. She is responsible for the effective competitive positioning of products, benefits, pricing and image, and directs a wide range of functional areas.

Dr. Alan Palkowitz is a drug discovery scientist who recently retired from Eli Lilly and Company, where he served for 28 years. His most recent leadership role was Vice President of Discovery Chemistry Research and Technologies. At Lilly, he supervised the work of more than 500 scientists and was responsible for small molecule (oral medicines) drug discovery for diseases that include cancer, diabetes, immunology, pain and neurodegenerative disorders.

Rajeev Ronanki is Senior Vice President and Chief Digital Officer of Anthem Inc., a Blue Cross Blue Shield company that is dedicated to providing consumers with affordable, high-quality health care. Rajeev has spent more than two decades in health care. Prior to joining Anthem in 2018, he spent a decade at Deloitte Consulting, where he led Deloitte's Cognitive Advantage (Markets and Technology) and Health Care Innovative practices. While there, he focused on implementing solutions for personalized consumer engagement, intelligent automation, and predictive analytics.

Amir Dan Rubin is President and CEO of One Medical Group, an innovative network of physicians and other care providers that offers its services at central locations in Boston, Chicago, Los Angeles, New York, the San Francisco Bay area, Seattle and Washington, D.C., with more locations to come.

Mario Schlosser is Founder and CEO of Oscar, a consumer-focused health insurer that is dedicated to using technology to humanize the experience of customer care. Oscar offers its members quick access to telehealth consultations with plan physicians, a dedicated healthcare concierge, a proprietary and full-featured health management app for all members, and a health clinic in Brooklyn that offers primary care along with wellness activities like yoga.

Travis Shannon is Vice President of Information at Leisure Sports Hospitality in Pleasanton, California, an industry leader in designing and managing upscale exercise and health clubs.

Constance Sjoquist is Chief Content Officer at HLTH.com, sponsors of the annual "HLTH: The Future of Healthcare" Conference. HLTH.com was founded by Jonathan Weiner, who is reshaping the industry narrative on what it means to drive reductions in health care costs, continually innovate and increase quality in health care.

Andy Slavitt is General Partner, Town Hall Ventures. During his career, Andy has led many initiatives that have shaped our national health care system and improved the lives of millions of Americans. Prior to launching Town Hall Ventures, he was Acting Administrator of the Center for Medicare and Medicaid Services (CMS) under President Obama.

Grant Verstandig is founder of Rally Health and Chief Digital Officer of the United Health Group. A young health care visionary to watch, Grant founded Rally Health in 2009. Today, Rally has joined forces with United Healthcare and now employs more than 3,000 people.

ABOUT THE AUTHOR

Kevin Pereau is Founder and Principal of TranscendIT Health, a boutique health care strategy and management-consulting firm focused on helping health care payers, providers and consumers get maximum value from digital health technologies.

A leading expert in the fields of digital and consumer health, Kevin spent his early career managing strategy and enterprise technology-consulting firms. The first health tech company he helped start was a consumer health scoring firm whose mission was to provide longitudinal context around how what we do and how we feel affects who we are.

Pereau frequently moderates panel discussions and speaks at health care conferences. He serves on the advisory boards of LifeData and HealthSaaS. He has advised a variety of digital health companies developing solutions focused on population health, telehealth, predictive analytics, marketplaces and consumer transparency tools.

Kevin grew up in a small town in Vermont, attended the University of Portland, OR, and now lives in Walnut Creek, CA, with his wife Beth, who is also a leading health care executive. Kevin's daughter Megan works as a health care consultant for MMA in Boston, MA.

9 780578 409726